VITAL
SIGNS
1997

VITAL SIGNS 1997

SIGNS

The Environmental Trends That Are Shaping Our Future

Lester R. Brown

Michael Renner

Christopher Flavin

Editor: Linda Starke

with

Janet N. Abramovitz

Seth Dunn

Hilary F. French

Gary Gardner

Hal Kane

Nicholas Lenssen

Anne McGinn

Jennifer D. Mitchell

Toni Nelson

Molly O'Meara

David M. Roodman

Kira Schmidt

Cheri Sugal

John Tuxill

W.W. Norton & Company

New York   London

The text of this book is composed in Garth Graphic
with the display set in Industria Alternate.

Composition by the Worldwatch Institute; manufacturing by the Haddon Craftsmen, Inc.
Book design by Charlotte Staub.

ISBN 0-393-31637-8 (pbk)

W.W. Norton & Company, Inc.
500 Fifth Avenue, New York, NY 10110
W.W. Norton & Company Ltd.
10 Coptic Street, London WC1A 1PU

 234567890

This book is printed on recycled paper.

AUGUSTANA UNIVERSITY COLLEGE
LIBRARY

Worldwatch Database Disk

The data from all graphs and tables contained in this book, as well as from those in all other Worldwatch publications of the past two years, are available on disk for use with IBM-compatible or Macintosh computers. This includes data from the State of the World *and* Vital Signs *series of books,* Worldwatch Papers, World Watch *magazine, and the* Environmental Alert *series of books. The data are formatted for use with spreadsheet software compatible with Lotus 1-2-3 version 2, including all Lotus spreadsheets, Quattro Pro, Excel, SuperCalc, and many others. For IBM-compatibles, a $3^1/_2$-inch (high-density) disk is provided. Information on how to order the Worldwatch Database Disk can be found on the final page of this book.*

CONTENTS

Part One: KEY INDICATORS

Part Two: SPECIAL FEATURES

ACKNOWLEDGMENTS

The group of individuals and organizations that made this sixth edition of *Vital Signs* possible extends far beyond the by-lines of each individual indicator or feature. The research and writing that produced this report received financial support from the W. Alton Jones Foundation and the U.N. Population Fund.

Because *Vital Signs* is an integral part of the Institute's overall work, much of the information contained in this book is an outgrowth of the research and information-gathering efforts that generate the annual *State of the World* report, Worldwatch Papers, and *World Watch* magazine. Therefore, our thanks are also due to the funders whose generous support makes these endeavors possible. They include the Nathan Cummings, Geraldine R. Dodge, Ford, William and Flora Hewlett, John D. and Catherine T. MacArthur, Edith Munson, Rasmussen, Turner, Wallace Genetic, Weeden, and Winslow foundations; the Foundation for Ecology and Development; The Pew Memorial Trust; the Rockefeller Brothers Fund; Rockefeller Financial Services; the Wallace Global Fund; and Robert Wallace, who made a personal contribution.

The *Vital Signs* series derives critical continuity and consistency from independent editor Linda Starke. Her meticulousness, eye for detail, and ability to craft 40 plus disparate individual drafts into a compatible set of indicators and features have left a distinct imprint on every edition of *Vital Signs* and of *State of the World* since their inceptions in 1992 and 1984, respectively.

This is the second year that we have produced *Vital Signs* with our in-house desktop publishing capacity. Aside from permitting us to make truly last-minute changes (and hence put out a more up-to-date book), this gives us greater control over design details and overall quality. Designer Elizabeth Doherty has made this a smooth and pleasant process. We also would like to acknowledge important production support from Lori Baldwin and Laura Malinowski. Lori is also in charge of putting together the Worldwatch Database Disk, which has all graphs and tables contained in Worldwatch publications.

The other behind-the-scenes participants in *Vital Signs* to whom we are indebted are Reah Janise Kauffman, who assists with fundraising that allows us to produce this volume each year; our administrative team of Suzanne Clift, Barbara Fallin, Bill Mansfield, Tara Patterson, and Amy Warehime; our communications team of Mary Caron, James Perry, and Denise Thomma; and our publications sales and support team of Millicent Johnson and Joseph Gravely. We are also grateful to Reah Janise and Bill for arranging for *Vital Signs* to be translated into some 17 languages.

With this edition of *Vital Signs*, we welcome staff researchers Seth Dunn, Jennifer Mitchell, Molly O'Meara, and Cheri Sugal as highly capable first-time contributors. We also had the benefit of several alumni—former staffers and interns—pitching in from a mix of new locations: Hal Kane in San Francisco; Nick Lenssen in Boulder, Colorado; Toni

Acknowledgments

Nelson in Washington, D.C.; globe-trotting
Kira Schmidt; and John Tuxill in Panama.

Authors received feedback on early drafts
and critical information from a variety of out-
side experts. We would like to thank Gerhard
Berz, Amie Brautigam, Mafa E. Chipeta,
Robert Davis, Mary Dickson, Jane Dignon,
Daniel Gallik, Paul Gipe, Joseph Grinblat,
Anna Gyorgy, James Hansen, Carl Haub,
Peter Johnson, Alan D. Lopez, Birger Madsen,
Paul Maycock, Christopher J.L. Murray,
Elliott Norse, Rachel Nugent, Mika Obayashi,
Enrique Panlo, Maurizio Perotti, Bill Quinby,
Annu Ratta, Knud Rehfeldt, Jac Smit, Daniel
Tarantola, Adam Tiller, Karen Treanton,
Timothy Whorf, and Brenda Lee Wilson.

Last but not least, we are once again grateful to Iva Ashner, Andrew Marasia, and their
colleagues at W.W. Norton & Company for
their support and ability to turn our camera-
ready manuscript into a published product at
a speed uncommon in the book-publishing
world.

Lester R. Brown
Michael Renner
Christopher Flavin

FOREWORD

We live in a world that is awash in statistics. Daily newspapers contain pages of financial information indicating how thousands of companies fared in the stock market the day before. Other pages are devoted to tracking mutual funds, currency exchange rates, and prices of commodities—from traditional items such as pork bellies to newer ones such as recycled paper. The political sections of papers are often laced with poll numbers, and even the sports pages are filled with columns of baseball box scores and team standings, so that busy readers will know "who is up and who is down" by the time they finish their morning coffees.

We are behaving somewhat like the fellow who looks only under the lamppost for keys lost in a vast parking lot—focusing not on the trends that are most important, but on those that are easiest to measure and report on, or that some government agency or industry has a particular interest in highlighting.

In doing so, we are missing a lot. Economic productivity is regularly reported, but the state of the natural world—whether it be the decline of coral reefs or the comeback of Zimbabwe's elephants—is rarely charted. The growth of world oil production is followed carefully by international agencies, but those same experts seem oblivious to the fastest-growing energy sources: wind and solar power. World population numbers are now reported by many news organizations, but a parallel decline in the number of human languages is not.

Since we first published *Vital Signs* in 1992, we have sought to broaden the base of information available to decisionmakers around the world by assembling a unique, eclectic set of global indicators. *Vital Signs* combines information on well-known trends such as grain production, population growth, and oil use with much rarer measures of the human condition. Although bicycle production, carbon emissions, and maternal mortality do not carry the same economic weight as traditional indicators, they too will deeply influence the lives of future generations.

In fact, *Vital Signs 1997* tracks a number of emerging trends that have the potential to shape the twenty-first century. The dramatic renaissance of the electric car and wind power industries in the 1990s may soon provide economical means of slowing global climate change. If they continue their spectacular double-digit growth rates, electric cars and wind turbines could displace enormous amounts of oil and coal. One day, wind energy may serve as the fuel for a new generation of electric cars.

We also see hope in *Vital Signs 1997* that humanity may be close to turning historic corners on some of its most pressing problems. Although human population stands at a record 5.8 billion, the annual additions to that number are now declining—with 80 million added in 1996, compared with 87 million in 1990. Efforts to provide family planning and health care and to raise the status of women are now paying off, though considerable additional progress is needed to stabilize world population.

Similarly, the overall HIV/AIDS infection rate continues to grow, but the rate of infection and the death rate are now slowing in some countries—thanks to a combination of public education and new drug therapies. The challenge now is to reach low-income groups with these solutions, particularly in the developing world.

The five earlier editions of *Vital Signs* have been published in 17 languages, and we have been gratified by the range of uses that readers have found for the report. Newspapers and magazines have reproduced hundreds of our graphs—sometimes on the business page, occasionally on the front page, and often in special features on particular topics.

Business executives have purchased the Institute's Database Disk, which contains all the numbers in *Vital Signs* and *State of the World* as well as in other Worldwatch publications, in order to incorporate the latest trends into their business plans. Government planners have used our figures in their official reports. And thousands of students have used *Vital Signs* as a source of facts for their term papers.

In future editions, we plan to continue adding new indicators to *Vital Signs*, and we will seek new ways to make the information even more accessible to users. Already, the Database Disk, first available in 1993, is proving popular with readers who want to perform their own analyses. We are hoping to upgrade it further in the near future.

Also, the Institute's Web Site (< http://www.worldwatch.org >) includes two sample *Vital Signs* as well as ordering information for *Vital Signs 1997* and other Worldwatch publications. At some point in the near future, we hope to make individual *Vital Signs* directly available to potential users over the Internet.

On behalf of our 14 coauthors, we want to thank you for your continued interest. Please let us know if you have ideas for improving *Vital Signs* or any other Worldwatch publications. We look forward to hearing from you.

Lester R. Brown
Michael Renner
Christopher Flavin
March 1997

Worldwatch Institute
1776 Massachusetts Ave., N.W.
Washington, DC 20036

VITAL
SIGNS
1997

OVERVIEW

A Year of Contrasts

Lester R. Brown

The year 1996 was one of sharp and sometimes disturbing contrasts. The production of chlorofluorocarbons (CFCs) fell, while carbon emissions continued their long-term rise. Economic growth in developing countries was almost triple that in industrial ones. Worldwide income per person climbed to a new high, while carryover stocks of grain dropped to an all-time low. Population growth slowed, but this was due to falling fertility in some countries and rising mortality in others.

The manufacture of CFCs, the family of chemicals that is depleting the stratospheric ozone layer that protects life on Earth from harmful ultraviolet radiation, has been declining for nearly a decade—setting the stage for the eventual healing of the ozone layer. Meanwhile, carbon emissions from fossil fuel burning climbed to a new high, laying the groundwork for a future with summers far hotter than any since agriculture began. In 1996, the Earth was only slightly cooler than in 1995, which was the warmest year since recordkeeping began in 1866.

This rise in carbon emissions has a contrasting economic relationship to two major industries. A rise in oil use in 1996 meant the oil industry prospered, yet the insurance industry suffered as weather-related disasters reached a record $60 billion. Some insurance industry leaders believe that the very reason for the prosperity of the oil companies—the growing use of oil and other fossil fuels, and the associated rise in atmospheric concentrations of carbon dioxide (CO_2), a greenhouse gas—may be responsible for the higher temperatures and greater storm intensity that is threatening the solvency of some companies. Accordingly, the chief executives of 60 of the world's leading insurance companies signed a statement urging governments to reduce carbon emissions and, hence, the use of fossil fuels.

In 1996, the contrast between the growth rates of industrial and developing countries increased as the growth rate of the latter averaged 6 percent—nearly three times the rate of industrial countries. As growth slows in more mature industrial economies and accelerates in developing countries, largely as a result of large transfers of private capital and technology, the stage is set for narrowing the gap between the have and have-not countries. In some areas, such as the use of fertilizer, developing countries have now moved ahead of industrial ones.

For the world as a whole, incomes moved up sharply in 1996, reaching an all-time high, while food security—measured in carryover stocks of grain—dropped to the lowest level

---

Units of measure throughout this book are metric unless common usage dictates otherwise. Historical population data used in per capita calculations are from the Center for International Research at the U.S. Bureau of the Census. Historical data series in *Vital Signs* are updated each year, incorporating any revisions done by originating organizations.

on record. One consequence of this growing imbalance between rising purchasing power and falling grain stocks was that prices of wheat and corn climbed to new highs in the spring of 1996.

The world population growth rate dropped to 1.4 percent in 1996, a decline from the historical high of 2.2 percent reached in 1963. Still, because of today's much larger population base, 80 million people were added to our numbers in 1996, compared with 69 million in 1963. As with many other trends discussed in this year's *Vital Signs*, the world is making headway on the population front, but the rate of progress will have to accelerate if we are to achieve a sustainable global economy for the next generation.

NEAR-RECORD ENERGY EXPANSION

Use of virtually every source of energy expanded in 1996. The fastest growing was wind energy, at 26 percent; the slowest was nuclear power, at just under 1 percent.

The three fossil fuels that account for 85 percent of the world's commercial energy use—oil, coal, and gas—expanded at 2.3, 1.8, and 4.5 percent, respectively. (See pages 46–47.) The growth in world oil consumption, now totalling 64 million barrels per day, is concentrated in Asia and North America. In both regions, increases in automobile and truck fleets accounted for much of the gain. With demand for oil running strong, prices rose above $20 a barrel, the highest since the Gulf War in 1991.

Trends in coal consumption were mixed, with coal use falling sharply in Western Europe and rising steadily in China. The growth of 4.5 percent in the use of coal in China, the world's leading user of this energy source, more than offset declines elsewhere, leading to the global rise.

Natural gas use is now rising throughout the world. Growth is particularly strong in Europe, where gas is replacing coal for both electricity generation and residential use. Among the fossil fuels, natural gas is most appealing because it is a cleaner burning fuel and yields fewer carbon emissions per unit of energy produced. Thus a shift from coal to gas helps industrial governments achieve the goal of stabilizing or reducing carbon emissions.

The use of geothermal energy expanded by 5.5 percent in 1996, boosting the total generating capacity of this energy source to 7,200 megawatts. (See pages 50–51.) Over half this capacity is concentrated in the United States, with 2,800 megawatts, and the Philippines, with 1,200 megawatts. Much of the remainder is scattered in some 20 other countries, mostly developing ones. Countries where geothermal power is growing rapidly include Indonesia, Japan, and China—all richly endowed with this renewable energy source.

The growth in wind power use in 1996 of 26 percent pushed its total generating capacity to nearly 6,100 megawatts. (See pages 52–53.) As in the preceding year, Germany led the world in new generating capacity, followed by India. Denmark, the world's leading supplier of wind turbines, ranked third in newly installed capacity. The manufacture of photovoltaic cells, the world's second fastest growing energy source, expanded by 16 percent in 1996. (See pages 54–55.) Installed capacity, which is now approaching 700 megawatts, is quite small by global standards, but this greatly understates the importance of this energy source simply because it supplies small amounts in strategic places. It is used, for example, in communication satellites to relay signals in the rapidly growing global electronic network. It powers most of the world's pocket calculators, and it is an increasingly popular source of energy in remote villages that are not linked to a grid. As of the end of 1996, some 400,000 homes were relying on electricity from solar cells.

CARBON EMISSIONS SET RECORD

Given the fossil fuel trends just described, it comes as no surprise that in 1996 the burning of oil, coal, and gas boosted carbon emissions to 6.25 billion tons, an all-time high. (See

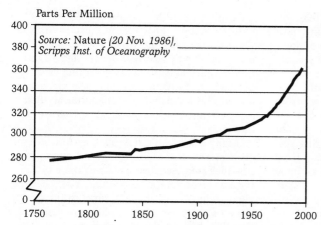

Parts Per Million

Source: Nature (20 Nov. 1986), Scripps Inst. of Oceanography

Figure 1: Atmospheric Concentration of Carbon Dioxide, 1764–1996

STORMS ROCK INSURANCE INDUSTRY

The Earth is getting warmer. Preliminary data show that 1996 was the fourth warmest year since recordkeeping began in 1866. (See pages 62–63 and Figure 2.) The 13 warmest years on record have occurred since 1979, with the four warmest being during the 1990s. The addition of the 1996 data provides yet another piece of evidence that a warming trend is under way.

The insurance industry is deeply concerned about this. Higher temperatures of surface waters, particularly in the tropics and subtropics, mean more heat is released into the atmosphere. As a result, storm systems are more intense, more frequent, and more destructive.

Damage from weather-related disasters in 1996 reached a record $60 billion. (See pages 70–71.) An estimated $26 billion of this occurred in China, where a succession of typhoons (hurricanes) led to severe flooding, displacing 2 million people and claiming 2,700 lives.

Weather-related insurance claims, a much

pages 58-59.) This 2.8-percent jump, the largest annual gain in nearly a decade, made the goal of stabilizing the Earth's climate seem even more unattainable.

The United States, the largest single source of carbon emissions, is responsible for 23 percent of the emissions of this climate-changing gas. China, the world's fastest growing economy during the 1990s, now accounts for 14 percent of carbon emissions, largely because of its heavy dependence on coal. Emissions there grew 27 percent from 1990 to 1995, compared with approximately 8-percent increases in both the United States and Japan in the same period.

Since the Industrial Revolution, atmospheric CO_2 levels have risen from an estimated 280 parts per million to 362 parts per million, the highest in 150,000 years. (See Figure 1.) The mainstream scientific community, represented by the Intergovernmental Panel on Climate Change—2,500 of the world's leading atmospheric scientists—now finds evidence that human activity is indeed altering the Earth's climate.

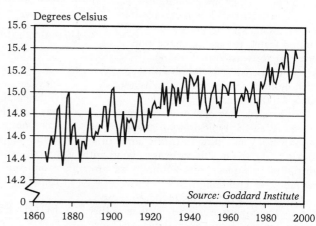

Degrees Celsius

Source: Goddard Institute

Figure 2: Average Temperature at the Earth's Surface, 1866–1996

smaller figure, totalled $9 billion in 1996, the third highest on record. Thus far during the 1990s, these claims have totalled $66 billion, compared with $17 billion during the 1980s.

Insurance companies are quite worried about the rising outlays associated with increasing weather damage, particularly from more destructive storms. As noted earlier, some 60 of the world's largest insurance companies signed a statement in 1996 urging governments to reduce emissions of carbon from fossil fuel burning. In addition, 13 large insurance companies combined resources to establish the Risk Prediction Initiative, located in Bermuda, to help forecast future climate trends.

BIKE OUTPUT TRIPLE THAT OF CARS

World automobile output totalled 36 million in 1996, enough to expand the global fleet to 496 million vehicles. (See pages 74-75 and Figure 3.) The largest growth occurred in Asia (excluding Japan), where the fleet expanded by 15 percent. As urban traffic congestion spreads and air pollution worsens, the social and environmental costs of growing reliance on the automobile are escalating, raising concerns within governments, both national and urban. In the United States, the traffic congestion price tag for wasted fuel, rising health care costs associated with air pollution, and lost productivity totals some $100 billion.

In Bangkok, one of the world's most congested cities, the typical motorist spends 44 days over a year sitting in traffic jams. In Mexico City, the health costs associated with air pollution, largely from automobiles, are estimated at $1.5 billion.

One response of governments to these automotive nightmares is to emphasize the use of bicycles. In 1995, the latest year for which reliable data are available, bicycle factories worldwide turned out an estimated 109

million bicycles. (See pages 76–77.) China, with an output of 41 million bicycles in 1995, was far and away the world leader. India, now in second place with more than 12 million bicycles assembled, is emerging as a bicycle power. Aside from the heavy reliance on bicycles for personal transportation, Asia is where most of the world's bicycles are produced.

Many European cities, such as Amsterdam, are fostering the use of bicycles. Copenhagen provides free bicycles for use in the city. In the European Union as a whole, bikes have been included, for the first time, in the comprehensive transportation plan. The United Kingdom has developed a plan to quadruple

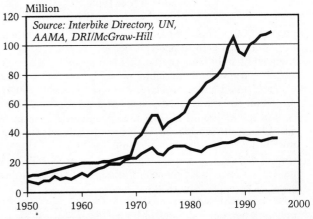

Figure 3: World Bicycle and Automobile Production, 1950–96

bicycle use by the year 2012. And Lima, Peru, is developing 51 kilometers of bikeways, and has plans to redesign 35 kilometers of roads to accommodate bicycles. This, along with a loan program to foster bicycle purchase among those with low incomes, is intended to avoid some of the acute traffic congestion found in other Third World cities.

FOOD SECURITY DETERIORATING

In 1996, world carryover stocks of grain dropped to 51 days of consumption, the low est level on record. (See pages 34–35.) In response, the intense competition among importing countries to get enough grain to make it to the next harvest drove wheat and corn prices to all-time highs in the spring of 1996—double the level of a year earlier.

This temporary doubling of prices reflects the loss of momentum during the 1990s in the growth of the world grain harvest. Among the contributing factors is the spreading scarcity of fresh water for irrigation; a diminishing response to the use of additional fertilizer in North America, Western Europe, the former Soviet Union, and Japan; and heavy losses of cropland to industrialization in Asia.

Meanwhile, growth in the world demand for grain is strong. The addition of some 80 million people each year generates a demand for 26 million more tons of grain. The rising affluence from rapid world economic growth during the mid-1990s is further bolstering growth in demand. In response to the tightening food situation, the United States has brought back into production all the land idled under its commodity supply management program. Europe has done the same with part of its set-aside land. With stocks down and little land remaining to be returned to production, the adequacy of future food supplies is now a matter of concern.

In 1996, the world grain harvest set an all-time record. (See pages 26–27.) Yet despite exceptionally favorable weather in major food-producing regions leading to the largest harvest ever, the world was unable to rebuild depleted stocks to a secure level. Carryover stocks are projected to increase from 51 days in 1996 to only 55 days in 1997, the second lowest level on record.

THE GROWING APPETITE FOR PROTEIN

One reason for the mounting pressure on the world's food-producing resources is the growing worldwide appetite for high-quality pro-

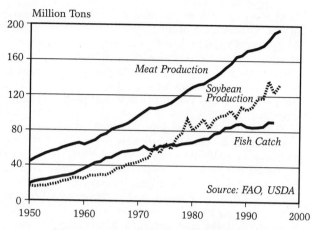

Figure 4: **World Protein Trends, 1950–96**

tein, including seafood, meat, and soybeans. (See Figure 4.)

The total fish catch increased from 19 million tons in 1950 to 89 million tons in 1989, but has gone up little since then. (See pages 32–33.) Even the 91-million-ton catch in 1995, the latest year for which data are available, may not be sustainable. With the wild catch levelling off, future demand for seafood can be satisfied only by expanding fish farming. Between 1985 and 1995, fish farm output more than doubled—from 8 million tons to 21 million tons. Continuing this expansion is technologically possible and economically feasible, but it puts mounting pressure on land and water resources because once fish are put in ponds or cages, they must be fed. The most common ration is corn and soybean meal, essentially the same as used in the poultry industry.

Between 1950 and 1996, the production of meat, the largest single source of animal protein, increased more than four times, from 44 million tons to 195 million tons. (See pages 30–31.) Of the four principal meats—beef, pork, poultry, and mutton—pork is the most widely consumed. Beef production has grown little in recent years. Poultry production, meanwhile, is soaring, nearly doubling over the last decade. Despite the doubling of corn, wheat, and barley prices in the spring of

1996, total world meat production continued to rise. Although the dramatic price rise brought growth in both pork and beef to a standstill, poultry production climbed by 6 percent. Overall, the jump in grain prices slowed growth in world meat production from 4 percent in 1994 and 1995 to less than 2 percent in 1996.

Demand for the third source of high quality protein in the world—soybeans—is growing even more rapidly. Between 1950 and 1996, the world soybean harvest increased from 17 million tons to 133 million tons, increasing nearly eightfold. (See pages 28–29.) Much of this was driven by the growing demand for soybean meal, the leading protein supplement in livestock and poultry feed. And part of it reflected climbing demand for vegetable oil in low-income countries where incomes are rising, such as China and India. At mid-century, soybeans were grown primarily because of the oil they contained. But in recent decades, as the need for a protein meal supplement in livestock rations has climbed, soybean meal has become the principal product, with oil as the by-product.

Worldwide, the direct consumption of soybeans for food is small, but it is growing rapidly. In China, for example, some 8.5 million tons of soybeans were consumed in 1996, mostly in the form of tofu, soy sauce, and bean sprouts. In western industrial countries, the consumption of tofu and of textured soybean protein as a meat substitute is expanding rapidly—for both economic and health reasons.

The extraordinary growth in protein consumption since mid-century reflects in part the increase in population, but it is also a measure of the rise in affluence. This occurred initially in western industrial countries, but now is concentrated in Asia.

ECONOMIC PACE PICKS UP

After a slow start in the early 1990s, the world economy has grown much faster in recent years. After increasing less than 4 percent from 1990 to 1993, the economy has expanded more than 11 percent during the last three years. (See pages 66–67.) In 1996, the world's output of goods and services rose by 3.8 percent, up slightly from the 3.5-percent growth in 1995.

Perhaps the most important contrast within the global economy in 1996 was between industrial and developing countries. While much attention had focused on the widening income gap between the world's poorest and its most affluent, the economies of the developing world grew an average of 6.3 percent in 1996, roughly three times as fast as industrial countries.

This growth is led by Asia, which contains more than half the world's people. Excluding Japan, the regional economy grew by roughly 8 percent in 1996 for the fifth consecutive year. China, which is not only the world's most populous country but also its fastest growing economy during the 1990s, expanded at an annual average of 10.1 percent during the decade's first six years.

The growth in the global output of goods and services from 1986 to 1996 totalled nearly $7 trillion. To put this in historical perspective, this exceeds the total growth in economic output from the beginning of agriculture until 1950.

The global economy is thus expanding at a robust rate. But the ecosystem on which it depends is not expanding at all. The evidence of this is painfully clear: the demand for seafood is exceeding the sustainable yield of fisheries; the grazing needs of the world's herds of cattle and flocks of sheep and goats are outpacing the sustainable yield of grasslands; forests are shrinking before the growth in the demand for firewood and lumber; aquifers are being depleted as the demand for water exceeds the recharge rate. And since much of this economic expansion is based on fossil fuels, carbon emissions into the atmosphere are outstripping the Earth's carbon-fixing capacity, leading to the buildup in greenhouse gases, rising temperatures, and growing climate instability described earlier.

Another sign that the economy is outgrowing the ecosystem is found in the decline in diversity of plant and animal species. Nowhere is this more evident than in the world's 232 nonhuman primate species. (See pages 100–01.) As a result of human population growth in the last few millennia, the number of humans now exceeds the total population of the other 232 primate species combined. And one fifth of these are threatened with extinction. As human numbers go up, the population of other primates goes down.

Other signs of excessive stress include the deterioration and death of coral reefs, the destruction of mangrove forests, and the draining of wetlands. (See pages 98–99.) Each of these ecosystems is an important building block of the global ecosystem on which all human economic activity depends.

POPULATION GROWTH SLOWING

In November 1996, U.N. demographers revised downward their estimates of population growth from 1985 onwards. Current figures show 41 million fewer people in the world in 1996 than earlier projections had indicated. The annual addition to world population fell from a peak of 87 million in 1990 to 80 million in 1996. (See pages 80–81.) In percentage terms, the annual addition peaked at 2.2 percent in 1963 and now stands at 1.4 percent.

The decline in population growth has come in part because of faster than projected declines in fertility in some key countries, such as India, Bangladesh, and Brazil. Worldwide, the fertility rate—the average number of children born to a woman in her lifetime—dropped from 4.2 in 1985 to 2.9 in 1996.

The bad news is that part of the recent slower growth in world population was due to rising mortality. In several African countries, newly available data on AIDS deaths suggest much slower population growth than had earlier been projected. Worldwide, an estimated 5.6 million new HIV cases—a record— brought the number of infections since the disease was first identified in 1982 to 36.2

million in 1996. (See pages 84–85.) The number of AIDS deaths climbed to 1.7 million, despite lower death rates in some industrial countries, where costly new treatments to control the disease are reducing fatalities.

Worldwide, more than half of those infected with HIV are in Africa. But the number of new cases in Asia in each of the last two years has exceeded the number in Africa. India, for example, now leads the world in HIV infections.

In the republics of the former Soviet Union, population growth is slowing because of rising mortality due to a breakdown in health care services, heavy smoking, and heavy consumption of alcohol. This translates into an increase in deaths from cardiovascular disease, accidents, murder, and suicides.

With life expectancy for men in Russia dropping from 64 in 1990 to 57 in 1995 and for women from 74 to 70, death rates are rising nationwide. Meanwhile, the birth rate has been falling, with the result that Russia's population is shrinking by nearly 1 million, or 0.6 percent a year—the fastest decline ever recorded in an industrial society.

WORLD IS DISARMING

The number of men and women in military uniforms is dropping steadily. In 1995, the last year for which data are available, it totalled 23 million, down 6 percent from the 24.6 million in 1994. (See pages 90–91.) The end of the cold war, the cessation of several small wars, and fiscal stringencies have all combined to reduce the number of troops by one fifth from the historical peak of 28.7 million in 1988. Among the countries that have made the heaviest cuts in the number of soldiers are China, Ethiopia, Iraq, Russia, Viet Nam, and the United States, where the number of Americans in the armed forces is at its lowest since 1950.

A similar trend has emerged with armaments. The number of battle tanks deployed worldwide declined from 172,000 in 1993 to 119,000 in 1996. (See pages 120–121.) Meanwhile, the number of combat aircraft dropped from 40,000 to 31,000. A combination of arms control agreements and peace treaties led to these reductions in arms deployments. Cuts were particularly large in countries where wars were ending, such as Angola, El Salvador, Ethiopia, Iraq, and Nicaragua.

Not surprisingly, arms manufacturing is also declining. Among the top 100 companies in the business outside China and the former Soviet Union, sales dropped from $186 billion in 1990 to $150 billion in 1994, the last year for which data are available. Trade in arms was in decline too, until 1995, when that trend was reversed.

For some countries, disarmament is not enough. Four countries now have no army at all: Costa Rica, which disbanded its army in 1949, has been joined by Iceland and most recently by Panama and Haiti, when the latter two overthrew military dictatorships. Barring a major international conflict or an increase in the number of civil wars, this arms reduction trend is likely to continue.

Part **ONE**

Key Indicators

Food

Trends

World Grain Harvest Sets Record   Lester R. Brown

A combination of strong prices at planting time, expanded area in grain, and unusually good weather helped make the 1996 grain harvest the largest ever: the record crop of 1.84 billion tons was 8.2 percent higher than the weather-depressed harvest of 1.70 billion tons in 1995.[1] (See Figure 1.)

That is the good news. The bad news is that there has been a dramatic loss of momentum in the growth in the world grain harvest during the 1990s. Between 1950 and 1990, the total harvested went from 631 million tons to 1,767 million tons, a gain of 180 percent—or 2.6 percent a year.[2] In contrast, between 1990 and 1996, the harvest went from 1.77 billion to 1.84 billion tons, an increase of only 4 percent or 0.7 percent a year.[3] This helps explain why carry-over stocks of grain in 1996 were at an all-time low and why prices of wheat and corn set record highs.[4]

Six years is not enough time to establish a new trend, but an analysis of the use of land, water, and fertilizer to produce grain gives cause for concern. The world grainland area in 1996, though up sharply from 1995, is less than it was in 1980.[5] The shrinkage can be traced to a combination of the conversion of cropland to nonfarm uses, particularly in Asia, where industrialization is moving at a record pace; the worldwide diversion of grainland to other crops, particularly soybeans; and the loss of land to degradation, mostly as a result of soil erosion.[6]

Harvested area expanded somewhat in 1996 as the modest amount of remaining set-aside land under commodity programs was returned to production in the United States and as the European Union reduced its set-aside area from 12 percent to 10 percent.[7] In Canada, China, Argentina, the European Union, and Russia, high grain prices led to the conversion to grain of some land that is normally planted to oilseeds.[8]

The growth in irrigated area is also slowing. Aquifer depletion, the diversion of water to cities, and the abandonment of irrigated area as the result of waterlogging and salting may now be offsetting any additions to the irrigated area. Despite official data showing some continuing growth, a 1996 report by the head of the International Irrigation Management Institute indicates that world irrigated area may now actually be shrinking.[9]

The diminishing response to the use of fertilizer in some countries where applications are heavy has actually led to a decline in usage in some situations. Growth in fertilizer use has levelled off or declined slightly in North America, Western Europe, and Japan.[10] In the former Soviet Union, market reforms that eliminated subsidies led to a precipitous drop in usage.[11]

The trend in per capita grain production clearly reflects this loss of momentum. Between 1950 and 1984, the world grain harvest per person went from 247 kilograms to 342 kilograms, a gain of 38 percent.[12] (See Figure 2.) After reaching this historical peak, it fell 2 percent by 1990, to 335 kilograms.[13] By 1996, it dropped a further 5 percent, to 319 kilograms per person.[14]

The 1996 rice harvest of 377 million tons set a new record—up nearly 2 percent from the previous record of 371 million tons in 1995.[15] And production of wheat, the other major food staple, totalled 579 million tons, 8 percent above the weather-depressed 1995 harvest of 537 million tons and second only to the 1990 harvest of 588 million tons.[16]

Corn production, meanwhile, at 573 million tons, was up 11 percent from the 514-million-ton harvest in 1995, which had been depressed by extreme heat in the U.S. Corn Belt, where more than one third of the world's corn is produced.[17] The 1996 harvest narrowly edged out the previous record of 560 million tons in 1994.[18]

The prospect for the 1997 world grain harvest, as always, will be influenced more by weather than by any other single factor. Prices are favorable. The European Union is reducing its set-aside area from 10 percent of its cropland in 1996 to 5 percent in 1997, a step that could add 2 million hectares of highly productive grainland to the area planted.[19] This gain could be offset, however, if strong prices in early 1997 for soybeans, sunflowers, rapeseed, and other oilseeds pull some grainland back into oilseeds.

WORLD GRAIN PRODUCTION, 1950–96

YEAR	TOTAL (mill. tons)	PER CAPITA (kilograms)
1950	631	247
1955	759	273
1960	824	271
1965	904	270
1966	989	290
1967	1,014	291
1968	1,053	296
1969	1,063	293
1970	1,079	291
1971	1,177	311
1972	1,141	296
1973	1,253	318
1974	1,204	300
1975	1,237	303
1976	1,342	323
1977	1,319	312
1978	1,446	336
1979	1,411	322
1980	1,429	321
1981	1,482	327
1982	1,533	332
1983	1,469	313
1984	1,632	342
1985	1,646	339
1986	1,663	337
1987	1,595	318
1988	1,548	303
1989	1,668	321
1990	1,767	335
1991	1,706	318
1992	1,786	328
1993	1,711	309
1994	1,759	314
1995	1,703	299
1996 (prel)	1,841	319

SOURCES: USDA, *Production, Supply, and Distribution* (electronic database), November 1996; USDA, "World Grain Database" (unpublished printout), 1991; USDA, FAS, *Grain: World Markets and Trade*, December 1996.

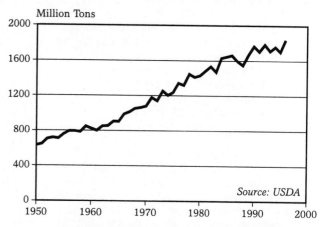

Figure 1: World Grain Production, 1950–96

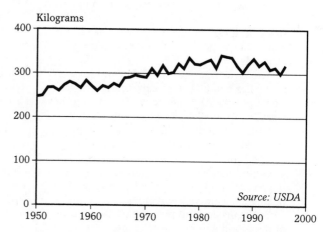

Figure 2: World Grain Production Per Person, 1950–96

Soybean Harvest Recovers to Near-Record   Lester R. Brown

In 1996, the world's farmers harvested 133 million tons of soybeans, up 7 percent from the weather-depressed 1995 harvest of 124 million tons.[1] (See Figure 1.) It was the second largest harvest on record, trailing only the 1994 harvest of 138 million tons.[2]

The soybean harvest per person totalled 23 kilograms in 1996, up from 22 kilograms in 1995.[3] (See Figure 2.) This was, however, still well below the historical high of 25 kilograms in 1994.[4]

Soybeans are unique among major crops in that their production is concentrated in a few countries: the United States, Brazil, Argentina, and China account for nearly 90 percent of the harvest.[5] A half-century ago, soybeans were still seen in the United States as a novelty crop introduced from China. By 1996, 49 percent of the world crop was grown in the United States, much of it produced in the Midwest in an alternate-year rotation with corn.[6] Farmers like this system because the soybean, a legume, fixes nitrogen for corn, a crop with particularly heavy nitrogen demands.

The growth in the world demand for soybeans has climbed at an extraordinary pace since mid-century. Between 1950, when the world's farmers harvested 17 million tons, and 1996, the harvest has expanded nearly eightfold.[7] This growth reflects the soaring demand for oilmeal as a supplement for livestock and poultry feed as well as the growth in consumption of vegetable oil among low-income consumers as incomes rise.

Soybeans account for roughly 30 percent of the world's vegetable oil supply.[8] The remaining dominant vegetable oils are peanut oil, sunflower seed oil, rapeseed oil, palm oil, coconut oil, and cottonseed oil.[9]

A small share of the world's soybean harvest—perhaps as much as one tenth of the total—is consumed directly as food.[10] In China, 8.45 million tons were eaten in 1996, mostly in the form of tofu, soy sauce, and bean sprouts.[11]

Most of the rest of the Chinese harvest—5 million tons or so—is crushed, yielding soybean oil for human consumption and oilmeal for use in livestock feed.[12] Like the demand for soybeans for food, that for both vegetable oil and oilmeals is climbing rapidly in China. It is estimated that roughly 10 percent of the mixed feed rations of 54 million tons produced in China in 1996 consisted of soybean meal, part of it imported.[13]

With soybeans, as with grain, China has recently been transformed from a net exporter into a net importer. In addition to importing unprecedented quantities of soybeans in the 1996/97 trade year, China is expected to bring in record amounts of both soybean oil and soybean meal.[14] With vegetable oil consumption rising some 70 percent from 1990 to 1995, China has emerged as the world's largest importer of vegetable oils.[15] Given the production declines in 1996, China's imports are expected to reach 3.45 million tons in 1996/97, up from 2.75 million tons in 1995/96.[16]

Soybean production is also growing in India, although this populous country is not yet one of the big four producers. India's 1996 harvest is estimated at a record 4 million tons.[17] As incomes rise there, the demand for vegetable oil is also increasing.

Growth in the demand for soybeans is expected to be strong in the years ahead. The demand in China just for food, for instance, is expected to increase from 8.45 million tons in 1996 to 13.8 million tons in 2000, a gain of more than 60 percent.[18] In other countries, such as the United States, consumption of soybeans as tofu and in the form of textured soy protein as a meat substitute is also expected to grow.

One reason for the growing popularity of soy-based foods in more affluent societies is the concern about the heavy intake of animal fats that can contribute to cardiovascular disease. Much of the future growth in demand is likely to be concentrated in Asia, where the consumption of vegetable oils and soy-based foods and the use of soymeal as a protein supplement in livestock and poultry feed are rising rapidly.

WORLD SOYBEAN PRODUCTION, 1950–96

YEAR	TOTAL (mill. tons)	PER CAPITA (kilograms)
1950	17	6
1955	19	7
1960	25	8
1965	32	9
1966	36	11
1967	38	11
1968	42	12
1969	42	12
1970	44	12
1971	47	12
1972	49	13
1973	62	16
1974	55	14
1975	66	16
1976	59	14
1977	72	17
1978	78	18
1979	94	21
1980	81	18
1981	86	19
1982	94	20
1983	83	18
1984	93	20
1985	97	20
1986	98	20
1987	104	21
1988	96	19
1989	107	21
1990	104	20
1991	107	20
1992	117	22
1993	118	21
1994	138	25
1995	124	22
1996 (prel)	133	23

SOURCES:: USDA, *Production, Supply, and Distribution*, electronic database, November 1996; USDA, FAS, *Oilseeds: World Markets and Trade*, December 1996.

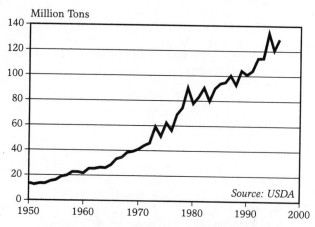

Figure 1: World Soybean Production, 1950–96

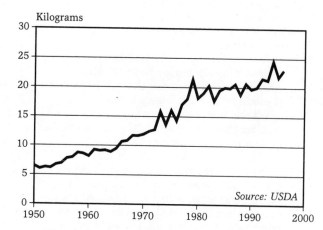

Figure 2: World Soybean Production Per Person, 1950–96

Meat Production Growth Slows   Lester R. Brown

World meat production in 1996 totalled 195 million tons, up from 192 million tons in 1995.[1] (See Figure 1.) This gain of 1.6 percent fell far short of the preceding year's near-record 4.3-percent gain, leaving meat consumption per person essentially unchanged.[2] (See Figure 2.)

Growth in meat output slowed in 1996 because of the unprecedented rise in feed-grain prices. Early in the year, prices of both corn and barley were more than double those at the beginning of 1995.[3] These record-high prices brought the growth in both beef and pork production to a standstill and slowed the growth in poultry from 8 percent in 1995 to 6 percent in 1996.[4] (See Figure 3.)

In addition to higher grain prices, the beef industry in Europe was hit by bovine spongiform encephalopathy (BSE), an infectious and incurable disease of cattle. This "mad cow" disease, concentrated in the United Kingdom, was linked in 1996 to an increasing incidence in humans of Creutzfeld-Jakob disease, a degenerative illness that affects the brain and central nervous system.[5] BSE has been traced to protein meals derived from the slaughter wastes of scrapie-infected sheep.[6] In an effort to eradicate it, the United Kingdom is slaughtering and incinerating 1.7 million older animals that were born before the use of these protein meals in feed was banned.[7]

The pattern of world meat production is shifting as beef production, which is largely grass-based, presses against the limits of the world's grazing area. In 1950, poultry production was scarcely a third that of beef; rapid growth since then pushed it upward, and in 1996 it overtook beef for the first time in history.[8] The gap between poultry and beef production is expected to widen in the years ahead.

The 6-percent growth in world poultry output in 1996 was concentrated in China, where production climbed by a phenomenal 18 percent, rising to 11 million tons.[9] This accounted for more than half the worldwide growth in poultry production. The United States, which boosted output from 13.8 million to 14.6 million tons, accounted for one fourth of the global increase.[10] Colombia,

France, and South Africa were responsible for much of the remainder.[11]

World poultry exports are also growing, passing 5 million tons in 1996 and apparently overtaking those of beef, which fell slightly.[12] China (including Hong Kong), Russia, and Japan account for almost three fourths of poultry imports.[13] The United States accounts for roughly half the exports.[14]

After expanding by 14 percent in 1995, the growth in pork production in China was stalled by rising grain prices.[15] As a result of higher feed prices, the world hog population is believed to have shrunk by 7 percent during 1996.[16] Cuts were largest in Mexico, Russia, Germany, and China.[17] Nevertheless, China still has just over half of the world's hog population and consumes half the pork.[18]

Growth in the production of beef, which has expanded slowly in the 1990s, ground to a halt in 1996 as grain prices climbed.[19] In the United States, which accounted for roughly one fourth of the output, production was up 2 percent, mainly because of the slaughter of herds by producers caught in the price squeeze.[20]

Beef production was down in the European Union largely because of the 14-percent drop in beef consumption following the BSE scare.[21] This decline, combined with that in the former Soviet Union, where the herd reductions of recent years continue, was offset by gains in Brazil, Canada, China, and, as noted, the United States.[22]

Beef imports are climbing, particularly in the densely populated countries of East Asia. Japan's beef imports in 1996 climbed to 960,000 tons, accounting for 62 percent of its consumption of 1.54 million tons.[23] Other land-scarce countries in Asia importing more beef include Taiwan, the Philippines, and Indonesia.[24]

U.S. Department of Agriculture analysts are projecting that the growth in world meat production in 1997 will be similar to that in 1996. Assuming lower grain prices in 1997, they are projecting no growth in pork production, a 1.5-percent growth in beef, and a 6-percent growth in poultry.[25]

WORLD MEAT PRODUCTION, 1950–96

YEAR	TOTAL (mill. tons)	PER CAPITA (kilograms)
1950	44	17.2
1955	58	20.7
1960	64	21.0
1965	81	24.2
1966	84	24.5
1967	86	24.5
1968	88	24.8
1969	92	25.4
1970	97	26.2
1971	101	26.7
1972	106	27.4
1973	105	26.8
1974	107	26.6
1975	109	26.6
1976	112	26.9
1977	117	27.6
1978	121	28.2
1979	126	28.8
1980	130	29.1
1981	132	29.2
1982	134	29.0
1983	138	29.4
1984	142	29.7
1985	146	30.1
1986	152	30.8
1987	157	31.2
1988	164	32.2
1989	166	32.0
1990	171	32.5
1991	173	32.2
1992	175	32.0
1993	177	32.1
1994	184	32.8
1995	192	33.7
1996 (prel)	195	33.8

SOURCES: FAO, *1948–1985 World Crop and Livestock Statistics* (1987); FAO, *FAO Production Yearbooks 1988–1991*; USDA, FAS, *Livestock and Poultry: World Markets and Trade*, October 1996.

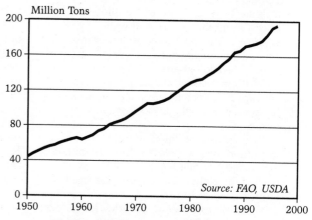

Figure 1: World Meat Production, 1950–96

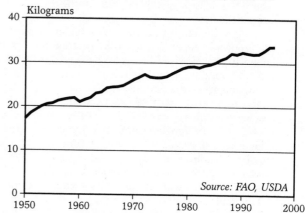

Figure 2: World Meat Production Per Person, 1950–96

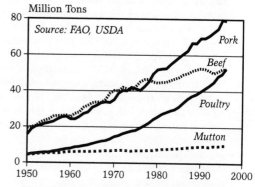

Figure 3: World Meat Production, by Type, 1950–96

Global Fish Catch Remains Steady   Anne McGinn

The total fish catch remained near record-high levels in 1995, with 91 million tons caught in marine and inland waters.[1] (See Figure 1.) The per capita catch declined to 15.9 kilograms, down by 7.5 percent from a peak of 17.2 kilograms in 1988.[2] (See Figure 2.) With increasing human population, the per capita fish catch is expected to decline further, even if wild stocks are allowed to recover.

Farmed fish, or aquaculture, increased to 21 million tons in 1995, up from 19 million tons in 1994.[3] (See Figure 3.) This growth pushed the global fish harvest—captured and farmed fish combined—to 112 million tons in 1995, an all-time record.[4] Worldwide, one out of every five fish eaten was raised on a farm, a share that is expected to increase in the years ahead.[5]

In 1994, the latest year for which regional data are available, an estimated 87 percent of the world's farm-raised fish were produced in Asia.[6] China dominated the region, as aquatic production grew at an annual rate of nearly 12 percent over the past 20 years.[7] China now harvests nearly one fourth of the world's fish overall.[8] In Bangladesh, China, India, Indonesia, and Thailand, cultured shrimp are one of the leading food exports.[9] Aquaculture in Asia as well as in Latin America is expected to continue growing in the years ahead.[10]

This rapid growth in fish farming carries with it high human and ecological costs. Shrimp aquaculture in Thailand, for instance, has been compared to "slash and burn" deforestation because of the environmental damage and pollution it causes.[11] Fish farms are essentially aquatic feedlots, requiring scarce land, water, feed, and energy inputs.[12]

Concerns such as these form the basis of a joint statement by leading nongovernmental organizations regarding unsustainable shrimp aquaculture that was presented at the May 1996 meeting of the U.N. Commission on Sustainable Development.[13] The statement argued that aquaculture is not a solution to the overexploitation of global fisheries, but that it can be promoted in a more environmentally and socially sustainable manner.

With more than 80 percent of the world's fish caught in the wild, farmed fish are only one part of the larger issue of overexploitation, however.[14] Japan's fishery output fell in 1995 for the seventh year in a row, due to large declines in the sardine and mackerel catch.[15] And in 1995 the U.S. National Marine Fisheries Service concluded that nearly half of all U.S. stocks were overexploited.[16]

One factor that may be affecting the long-term productivity of wild stocks is trawling. Recent studies by marine biologists highlight the destruction caused by fishing trawlers, draggers, and other mobile gear that scrape the ocean floor and disrupt ecosystems while scooping up fish and marine life.[17]

One study estimated that for every kilogram of commercial catch, between 9 and 18 kilograms of sea urchins, sponges, and other deep sea species are caught.[18] Although some of these species may not be commercially viable, many are an important food source for larger species and play a vital role in the marine ecosystem. These studies raise troubling questions about the extent to which commercial fish stocks depend on habitats that are being degraded by trawling, coastal aquaculture, and other damaging fishing methods.

In an effort to curb excess capacity, the European Union (EU) proposed to set aside $2.2 billion to fund restructuring and economic transition programs for fishers.[19] But intense pressure from industry officials, politicians, and fishers prompted EU ministers to minimize fleet reductions and to postpone indefinitely some desperately needed economic adjustments.[20]

If current trends of overexploitation and habitat destruction continue, fish will no longer be "the protein of the poor." And fishing will cease to be the primary source of income for the more than 200 million people who are directly involved in it worldwide.[21]

WORLD FISH CATCH, 1950–95

YEAR	TOTAL (mill. tons)	PER CAPITA (kilograms)
1950	19.2	7.5
1955	26.4	9.5
1960	36.4	12.0
1965	49.0	14.7
1966	52.6	15.4
1967	55.6	16.0
1968	56.5	15.9
1969	57.4	15.8
1970	58.2	15.7
1971	62.4	16.5
1972	58.4	15.1
1973	59.0	15.0
1974	62.6	15.6
1975	62.4	15.3
1976	64.6	15.5
1977	63.4	15.0
1978	65.3	15.2
1979	66.1	15.1
1980	67.0	15.0
1981	69.4	15.3
1982	71.1	15.4
1983	71.6	15.3
1984	77.0	16.1
1985	78.6	16.2
1986	84.0	17.0
1987	84.5	16.8
1988	88.0	17.2
1989	88.7	17.1
1990	85.4	16.2
1991	84.6	15.8
1992	84.9	15.6
1993	85.7	15.5
1994	91.0	16.2
1995	90.7	15.9

SOURCES: FAO, *Yearbook of Fishery Statistics: Catches and Landings* (Rome: various years); 1990–95 data from FAO, Rome, letter to author, 8 November 1996.

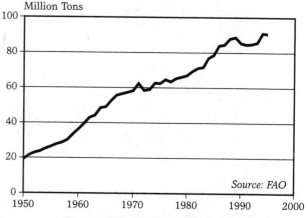

Figure 1: World Fish Catch, 1950–95

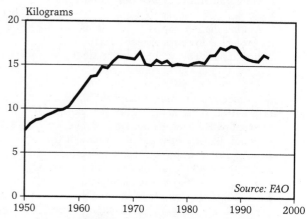

Figure 2: World Fish Catch Per Person, 1950–95

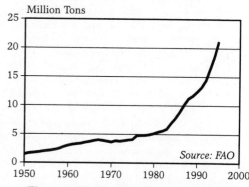

Figure 3: World Aquaculture Production, 1950–95

Grain Stocks Up Slightly

Lester R. Brown

Carryover stocks of grain in 1997—that is, the amount left in the bins from 1996 when the new harvest begins—are projected to total 272 million tons, or 55 days of consumption.[1] (See Figures 1 and 2.) This is up from 246 million tons in 1996 (the equivalent of 51 days of consumption), which was the lowest on record.[2]

After three consecutive years in which world grain consumption exceeded production, a combination of strong prices, expanded area in grain, and good weather boosted the 1996 harvest above consumption, although only marginally.[3] Having just 55 days of carryover stocks is still dangerously low. Whenever grain stocks drop below 60 days of consumption, grain prices become volatile, vulnerable to weather-reduced harvests.

Among the individual grains, stocks of rice are projected to remain steady in 1997, at 51 million tons.[4] This will leave rice stocks at only 50 days of world consumption.[5]

The world rice harvest climbed to a new high for the third consecutive year.[6] Though its growth is remarkably stable, it is not keeping pace with increased demand. If the excess of demand over supply in four of the last six years continues, further reducing stocks, this could lead to a dramatic rise in price as competition for inadequate supplies among importing countries intensifies.

For wheat, carryover stocks in 1997 are estimated at 112 million tons, the equivalent of 72 days of use.[7] This is far higher than the stocks of rice, but it is needed because so much of the world's wheat is grown in marginal rainfall areas and is thus vulnerable to even modest declines in rainfall.

Carryover stocks of coarse grains (largely corn, but including small amounts of barley, sorghum, millet, oats, and rye) are projected at 109 million tons in 1997, up from just over 90 million tons in 1996.[8] This will equal 46 days of consumption, compared with 39 days in 1996—the lowest level on record.[9] Since coarse grains are used largely for livestock and poultry feed, having low carryover stocks is not as life-threatening as it is for rice and wheat, humanity's dominant food staples.

In many ways, carryover stocks of grain are the most sensitive food security indicator. Having 70 days' worth is considered desirable for a minimum level of food security. This is enough to absorb most of the destabilizing effects of one poor grain harvest. With anything less, the world is at risk, particularly since the United States—which accounts for nearly half of world grain exports—relies heavily on rainfall to produce its grain crop.[10] This contrasts sharply with China, where two thirds or more of the grain harvest comes from irrigated land.[11]

Weather has always affected food security, but now climate change could be affecting the harvest prospect. The 13 warmest years since recordkeeping began in 1866 have all occurred since 1979.[12] And the four warmest years within the 13 are in the 1990s.[13] With rising temperatures, the United States is particularly vulnerable to crop-withering heat waves of the sort that reduced its grain harvest in 1988, 1991, and 1995—three of the last nine years.[14] If the rise in global average temperature that has been under way since 1979 continues, crop-damaging heat waves could become both more severe and more frequent.

The world's farmers are struggling to feed more than 80 million more people each year, good weather or bad.[15] And now, for the first time in history, they can no longer count on fishing fleets to help them expand the food supply. The combination of record low grain stocks, the rise in average global temperature since 1979, and the inability of fishers to help feed a growing population all suggest that maintaining an acceptable level of food security may be far more difficult in the future.

WORLD GRAIN CARRYOVER STOCKS, 1961–97[1]

YEAR	STOCKS (mill. tons)	(days use)
1961	203	90
1962	182	81
1963	190	82
1964	193	83
1965	194	78
1966	159	62
1967	189	72
1968	213	78
1969	244	87
1970	228	77
1971	193	63
1972	217	69
1973	180	56
1974	192	56
1975	200	61
1976	220	66
1977	280	80
1978	279	77
1979	328	86
1980	316	81
1981	289	72
1982	308	77
1983	357	88
1984	305	73
1985	366	85
1986	434	100
1987	465	104
1988	404	89
1989	314	70
1990	295	64
1991	339	72
1992	324	69
1993	363	76
1994	317	66
1995	302	62
1996	246	51
1997 (prel)	272	55

[1]Data are for year when new harvest begins.
SOURCES: USDA, FAS, *Grain: World Markets and Trade,* January 1997; USDA, *Production, Supply, and Distribution,* electronic database, November 1996.

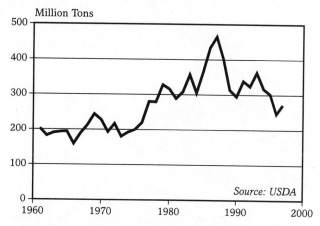

Figure 1: World Grain Carryover Stocks, 1961–97

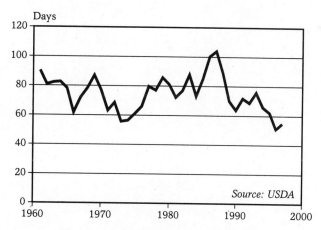

Figure 2: World Grain Carryover Stocks as Days of Consumption, 1961–97

Agricultural Resource
Trends

Fertilizer Use Rising Again
Lester R. Brown

World fertilizer use in 1996 climbed to 128 million tons, up 5 percent from the 122 million tons used in 1995.[1] (See Figure 1.) This marked the second annual increase after five years of consecutive decline.

Most of the growth in 1996 was concentrated in China. (See Figure 2.) Higher grain prices accounted for much of the 3.8-million-ton growth in fertilizer use by Chinese farmers in 1996.[2] This pushed China's fertilizer applications above 32 million tons, the highest ever for any country.[3] In the United States, fertilizer use increased from 20 million to 21 million tons.[4] This was largely due to the return to production of some 7.5 percent of cornland that had been set aside in 1995.[5]

Another region of growth was the Indian subcontinent, where fertilizer use climbed from 17.1 million to 17.8 million tons—up more than 4 percent.[6] In contrast to many regions where fertilizer use has fluctuated from year to year, growth has been remarkably steady in India, Pakistan, and Bangladesh.

In Western Europe, a modest reduction of set-aside land in 1996 would normally have led to an increase in fertilizer use, but adverse weather in some countries and environmental restrictions on fertilizer use in others held it steady.[7]

In Africa, a region that used only 2.5 million tons of fertilizer in 1996, there was no perceptible change in application rates.[8] In Latin America, meanwhile, despite a big gain in Argentina, fertilizer use actually declined, largely because of adverse economic conditions in Mexico and Brazil.[9]

From 1950 to 1989, world fertilizer use climbed to a new high almost every year.[10] Higher-yielding crop varieties that were more responsive to fertilizer and a steady expansion in irrigation were responsible for most of this growth. Throughout most of this period, more fertilizer meant higher yields. But by 1990 this was beginning to change.

For U.S. farmers, application rates had reached a level where additional fertilizer use had little effect on production. As a result, usage levelled off or declined a bit. In the mid-1990s, U.S. farmers are using slightly less fertilizer than in the early 1980s.[11] (See Figure 3.) A similar trend unfolded at about the same time in Western Europe and Japan.[12]

Fertilizer use in the former Soviet Union continued climbing during the 1980s as central planners in Moscow tried desperately to eliminate their dependence on grain imports.[13] They subsidized heavily the use of fertilizer, leading to excessive use.

As the agricultural reforms that were adopted in 1988 shifted the economy toward a market-based system, fertilizer prices climbed sharply.[14] With the breakup of the former Soviet Union and a severe economic depression, fertilizer use fell from a peak of just over 27 million tons in 1987 to scarcely 4 million tons in 1995.[15] In 1996, fertilizer use increased slightly, suggesting not only that the decline has ended, but that an agricultural recovery may be under way.[16]

Between 1950 and 1990, world irrigated area expanded from 94 million hectares to 240 million hectares.[17] Since then, it has increased very little. If it no longer expands, as some analysts expect, this will constrain future growth in fertilizer use.[18] To use fertilizer effectively, farmers need an abundance of soil moisture, either from natural rainfall, as in the U.S. Midwest or Western Europe, or from irrigation, as in China, where roughly half of the cropland is irrigated.[19]

Fertilizer is used on all crops, but about 60 percent is used to produce grain—both food grains and feedgrains.[20] Some of the more commercialized crops, such as oilseeds, fibers, sugars, and beverages, account for roughly 18 percent of world fertilizer use.[21] The remaining 22 percent is applied to food crops—such as fruits and vegetables, pulses, and roots and tubers—and to the fodder and pasture used to produce meat and milk.[22]

Given the low level of world grain stocks in 1996 and the projected continuing low level in 1997, grain prices are likely to remain strong.[23] This will encourage more fertilizer use where there is still a substantial agronomic potential for increasing production, including in Argentina, Myanmar, the Ukraine, and the Indian subcontinent.

WORLD FERTILIZER USE, 1950–96

YEAR	TOTAL (mill. tons)	PER CAPITA (kilograms)
1950	14	5.5
1955	18	6.5
1960	27	8.9
1965	40	12.0
1966	45	13.2
1967	51	14.6
1968	56	15.7
1969	60	16.5
1970	66	17.8
1971	69	18.2
1972	73	18.9
1973	79	20.1
1974	85	21.2
1975	82	20.1
1976	90	21.6
1977	95	22.4
1978	100	23.2
1979	111	25.3
1980	112	25.1
1981	117	25.8
1982	115	24.9
1983	115	24.5
1984	126	26.4
1985	131	27.0
1986	129	26.1
1987	132	26.3
1988	140	27.4
1989	146	28.1
1990	143	27.1
1991	138	25.7
1992	134	24.6
1993	126	22.8
1994	121	21.6
1995	122	21.4
1996 (prel)	128	22.2

SOURCES: FAO, *Fertilizer Yearbook* (various years); International Fertilizer Industry Association, Annual Conference, 19–22 November 1996.

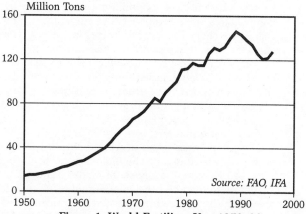

Figure 1: World Fertilizer Use, 1950–96

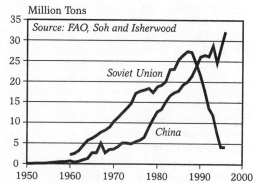

Figure 2: Fertilizer Use in China and the Soviet Union, 1950–96

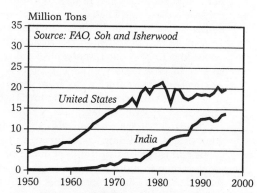

Figure 3: Fertilizer Use in the United States and India, 1950–96

Grain Area Jumps Sharply   Gary Gardner

The world's grain harvested area jumped 2.5 percent in 1996, from 679 million to 696 million hectares.[1] (See Figure 1.) The surge, triggered by a sharp rise in grain prices in 1995 and 1996, was the largest percentage increase in grain area since 1975.[2] The expansion even outpaced population growth—a rare occurrence since mid-century—thereby boosting grain area per person for the first time since the mid-1970s.[3] (See Figure 2.)

Grain harvested area is the grain acreage reaped in one year. (Land harvested twice a year is double-counted.) The indicator tracks the resource base of a key set of foods: grains supply more than half the calories and protein eaten directly by humans.[4] Major grains include corn, the area of which expanded by 5 percent in 1996; wheat, which increased by 1.4 percent; and rice, which was virtually unchanged.[5]

The 1996 jump in area is a deviation from the steady contraction since 1981.[6] In that year, global grain harvested area peaked at 732 million hectares after a steady rise since mid-century.[7] The reduction in area through 1995 put great pressure on yields to meet annual increases in demand.[8]

Several factors influence the amount of farmland planted to grains. Government set-aside programs, which idle land in order to minimize surpluses, have long restricted the amount of land in production. But this is changing. In the United States, the 1996 Farm Bill ended set-asides except for environmental purposes; these had idled an average of 13 million hectares annually between 1986 and 1995.[9] The European Union has reduced its set-aside acreage from 12 percent of cultivated area in 1995 to only 5 percent for 1997.[10] As governments venture away from set-asides, their countries are moving toward full capacity in agricultural land use.

The extent of grain area is also affected by grain prices. Declining prices through the 1980s and early 1990s prompted many farmers to switch to more lucrative crops.[11] Land planted to nongrains is sometimes described as "reserve" grainland because it can be tapped again for grain production, as some

was in 1996.[12] Oilseed area, for example, fell by 4 million hectares in 1996 as oilseeds were replaced by grains.[13] But if demand for soybeans, fruits, and vegetables remains strong, these crops may increasingly compete with grains for available land.[14]

Damage to agricultural land, especially from erosion and salinization, also pulls land from production, and may be on the increase: while losses of degraded land between 1945 and 1990 averaged some 2 million hectares annually, various sources suggest that 5–10 million hectares are now ruined each year.[15]

Countries that once relied on centrally planned economies have seen especially large losses of grain area. In the former Soviet Union, for example, grain area fell from 123 million hectares in 1979 to 91 million in 1995.[16] Some losses stem from disruptions in the flow of inputs and credits caused by the transition to market economies. But in many cases the land is simply exhausted. Kazakstan, for example, has retired more than 6 million hectares of marginal grainland since 1987; yields on this land averaged only a third of world levels.[17]

Grainland is also lost to urban expansion. Rapid economic growth in Asia (outside Japan)—averaging 8 percent annually since 1992—is devouring grainland as houses, factories, roads, and parks multiply.[18] China lost 3 percent of its cropland between 1986 and 1992; Vietnam, Indonesia, and Malaysia have also seen severe losses.[19]

The continuing loss and degradation of grainland and the lack of sustainably farmable virgin territory in most of the world has forced some poor farmers onto marginal land, much of it forested. Agricultural expansion is a leading cause of deforestation globally; the World Bank warned in 1996 that farm expansions could destroy nearly half the world's tropical forests.[20]

By protecting the world's remaining sustainably cultivable grain areas, governments can minimize the environmental damage caused by imprudent agricultural expansion, and can preserve the world's capacity to produce staple foods.

WORLD GRAIN HARVESTED AREA, 1950–96

YEAR	TOTAL (mill. hectares)	PER CAPITA (hectares)
1950	587	0.23
1955	639	0.23
1960	639	0.21
1965	653	0.20
1966	655	0.19
1967	665	0.19
1968	670	0.19
1969	672	0.18
1970	663	0.18
1971	672	0.18
1972	661	0.17
1973	688	0.17
1974	691	0.17
1975	708	0.17
1976	717	0.17
1977	714	0.17
1978	713	0.17
1979	711	0.16
1980	722	0.16
1981	732	0.16
1982	716	0.16
1983	707	0.15
1984	710	0.15
1985	715	0.15
1986	709	0.14
1987	685	0.14
1988	687	0.13
1989	693	0.13
1990	693	0.13
1991	690	0.13
1992	692	0.13
1993	683	0.12
1994	684	0.12
1995	679	0.12
1996 (prel)	696	0.12

SOURCES: USDA, *Production, Supply, and Distribution,* electronic database, February 1996; USDA, *Grain: World Markets and Trade,* January 1997.

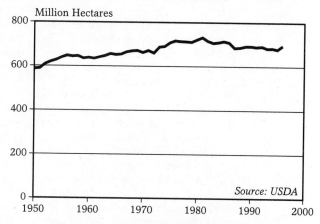

Figure 1: World Grain Harvested Area, 1950–96

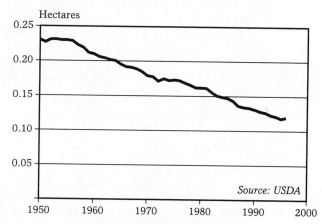

Figure 2: World Grain Harvested Area Per Person, 1950–96

Irrigated Area Up Slightly   Gary Gardner

World irrigated area increased from 247 million to 249 million hectares in 1994, the last year for which global data are available.[1] (See Figure 1.) The sluggish advance—just 0.7 percent—lagged behind population growth, thereby shrinking irrigated area per person.[2] (See Figure 2.) At just over 44 hectares per thousand people, per capita area has dropped to its lowest level since the 1950s.[3]

The slowing growth in irrigated area is part of an established trend. Annual irrigation expansion fell from more than 2 percent in the mid-1970s to less than 1 percent in the first half of the 1990s.[4] Area per person peaked in 1978 and declined in 12 of the next 16 years.[5]

The trend in irrigation varies by region. Developing countries have shown the highest rates of growth in the 1990s, ranging from 1.3 percent annually in Asia to 2.2 percent in Latin America and 2.7 percent in Africa.[6] Only in Latin America, however, did irrigation growth outpace the increase in population.[7]

Meanwhile, irrigated acreage fell more than 13 percent in Eastern Europe and more than 7 percent in the former Soviet Union between 1989 and 1994 as these regions shifted away from command economies.[8] To the extent that these declines stem from economic disruption and depression, the lost area may one day be recovered.

Sluggish growth in global irrigated area is worrisome because of irrigation's contribution to agricultural productivity. By providing water in a timely and measured manner, irrigation can make two or even three crops possible on the same land each year.[9] As a result, irrigated areas, although only 17 percent of the global cropland, produce 40 percent of the world's food.[10] This productivity, combined with expansion of overall area, made irrigation responsible for more than half of the increase in global food production between the mid-1960s and the mid-1980s.[11]

Several factors account for the slowing growth in irrigation and for the poor prospects for expansion. In many areas, water is simply unavailable for additional irrigation. Some 44 countries—most in Africa and the Middle East—are now "water-stressed," with annual renewable supplies of 1,700 cubic meters per person or less; most cannot meet the full water needs of farmers, factories, homes, and the natural environment.[12] When cuts in deliveries come, farmers are typically hit the hardest.

Even if water is available, irrigation projects are increasingly expensive and unpopular. The most affordable projects are already built, and most irrigation investment today is used to rehabilitate or maintain dams and canals.[13] Social and environmental costs can also be high: dams displace growing numbers of people and can disrupt river ecosystems above and below the dam site.[14]

An estimated 2 million hectares of irrigated land are pulled from production each year because of waterlogging and salinization, the result of poor irrigation management.[15] In addition, irrigated land is paved over for housing, factories, and roads, most notably in Asia and the United States.[16] California, for example, saw irrigated area fall by some 25,000 hectares—nearly 1 percent of its irrigated area—just between 1990 and 1992.[17] Continued losses at this rate would cost the state nearly 9 percent of its irrigated area in 25 years.[18]

In addition, some cropland is unsustainably irrigated. In northern China, northern India, Libya, Iran, the Arabian peninsula, and parts of the United States, millions of hectares are watered by pumping aquifers faster than rainfall can replenish them.[19] These regions may see cutbacks in irrigated area as falling water tables increase the cost of pumping, or as aquifers are depleted or polluted. On the Texas High Plains, for example, irrigated area fell by more than 25 percent between 1974 and 1989, in part because of overpumping, and despite increasing efficiency of water use.[20]

For these reasons, the prospects for irrigation expansion are bleak. One study projects growth of 0.3 percent annually over the next 50 years, less than half the anemic 1994 rate.[21] If true, the burgeoning global demand for food will be met only through a dramatic boost in water use efficiency.

WORLD IRRIGATED AREA, 1961–94

YEAR	TOTAL (mill. hectares)	PER CAPITA (hectares per thousand population)
1961	139	45.1
1962	141	45.1
1963	144	45.0
1964	147	44.8
1965	150	44.8
1966	153	44.8
1967	156	44.8
1968	159	44.8
1969	164	45.1
1970	167	45.1
1971	171	45.1
1972	174	45.1
1973	180	45.6
1974	183	45.7
1975	189	46.1
1976	194	46.6
1977	198	46.8
1978	204	47.3
1979	207	47.2
1980	209	46.9
1981	213	46.9
1982	214	46.4
1983	216	45.9
1984	221	46.2
1985	223	46.0
1986	225	45.6
1987	227	45.1
1988	230	44.9
1989	235	45.2
1990	240	45.3
1991	242	45.0
1992	245	44.8
1993	247	44.5
1994	249	44.1

SOURCES: FAO, *Production Yearbook* (various years); Bill Quinby, USDA, ERS, letter to author, 24 January 1996.

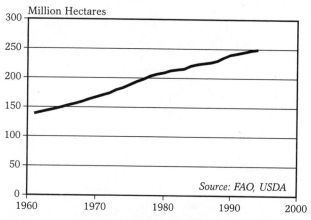

Figure 1: World Irrigated Area, 1961–94

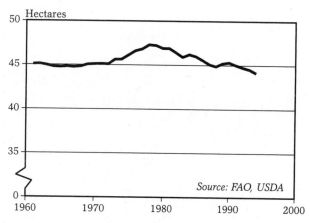

Figure 2: World Irrigated Area Per Thousand People, 1961–94

Energy
Trends

Fossil Fuel Use Surges to New High   Christopher Flavin

In 1996, dependence on fossil fuels grew to the highest levels ever, as use of oil, natural gas, and coal all set new records.[1] (See Figure 1.) At 8.1 billion tons of oil equivalent in 1996, fossil fuels provided roughly 85 percent of the world's commercial energy.[2] Rapid economic growth and moderate fuel prices have spurred the recent increases. Also, after falling steeply for several years, fossil fuel use in the former Soviet Union and central Europe is now levelling off, which has allowed the global totals to jump upward.[3]

Use of oil, the dominant fossil fuel, grew by an estimated 2.3 percent in 1996—to hit an all-time high of 3.3 billion tons (64 million barrels per day).[4] (See Figure 2.) Rising demand for oil in 1996 helped push prices above $20 a barrel for the first time since the Gulf War in 1991.[5] The fastest growth in oil use occurred in Asia and North America, while use fell 4.6 percent in the former Eastern bloc—the sixth annual decline in that region.[6] Throughout the world, increased reliance on motor vehicles is fueling the record demand for oil.

Production of oil also set new highs in 1996, surpassing records set in the late 1970s.[7] Although nearly two thirds of the world's remaining oil reserves are in the Middle East (with one quarter in Saudi Arabia alone), production there increased only 1 percent in 1996.[8] But production continued to soar in the North Sea, which has driven West European output up by a remarkable 56 percent since oil prices collapsed in 1986.[9] Oil production rose 4.8 percent in Africa in 1996, while in Asia it rose 2.1 percent. However, output fell 1.3 percent in the United States—where it is now 30 percent below the peak of the early 1970s.[10]

Analysts believe that global oil production could rise another 20–30 percent before peaking sometime between 2005 and 2015.[11] Major, sustained price increases appear unlikely in the next several years—barring a political cataclysm in the Middle East.

The use of coal increased 1.8 percent in 1996, but is still only 1 percent above the 1989 level.[12] Coal use declined sharply in Western Europe in 1996 for the ninth year in a row.[13] In Germany and the United Kingdom, the consumption of coal is now more than one third below the peak levels of the mid-1980s.[14] It declined in Russia and central Europe as well in 1996, as heavy industries continued to close and others shifted to natural gas.[15]

Offsetting the declines in Europe was increased use of coal in the United States in 1996—by a startling 3.7 percent—driven in part by cold winter weather and high prices for natural gas.[16] Coal use is rising even faster in China, the world's leading coal producer and consumer—up 4.5 percent in 1996, and 50 percent over the last decade.[17] China's heavily subsidized coal industry employs 3.5 million people, but as the subsidies are withdrawn, Coal Minister Wang Senhao expects production to stagnate in the late 1990s.[18]

World use of natural gas, the least environmentally damaging of the fossil fuels, rose the fastest in 1996—up 4.5 percent from the year before.[19] This is the largest increase in gas consumption since 1988, and appears to portend a period of accelerated growth.[20] New technology and the opening of new markets in regions such as Asia and the Middle East will allow gas to substitute for coal and oil in many applications.

The previous year's decline in natural gas use in Russia reversed course in 1996, pushing global gas totals upward.[21] And Russia is stepping up efforts to export its gargantuan gas reserves to neighboring countries, including China and Turkey.[22] China, which gets a meager 2 percent of its current energy from natural gas, is also expanding domestic gas exploration.[23]

The 10-percent boom in gas use in the relatively mature energy markets of Western Europe was notable in 1996, as demand for clean fuels grew, North Sea gas production expanded, and fear of reliance on Russian gas subsided.[24] European gas use is likely to expand further in the years ahead as its gas markets are opened to competition for the first time.[25]

WORLD FOSSIL FUEL USE, 1950–96

YEAR	USE (mill. tons of oil equivalent)
1950	1,715
1955	2,230
1960	3,019
1965	3,746
1966	3,949
1967	4,018
1968	4,330
1969	4,651
1970	4,744
1971	5,158
1972	5,408
1973	5,487
1974	5,754
1975	5,456
1976	5,755
1977	5,951
1978	6,150
1979	6,331
1980	6,343
1981	6,269
1982	6,213
1983	6,268
1984	6,512
1985	6,724
1986	6,905
1987	7,103
1988	7,346
1989	7,529
1990	7,502
1991	7,472
1992	7,504
1993	7,543
1994	7,727
1995	7,856
1996 (prel)	8,076

SOURCE: Worldwatch estimates based on DOE, BP, UN, and government sources.

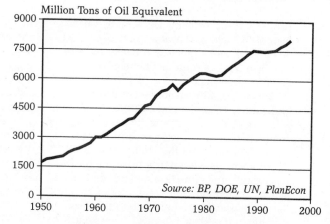

Figure 1: World Fossil Fuel Use, 1950–96

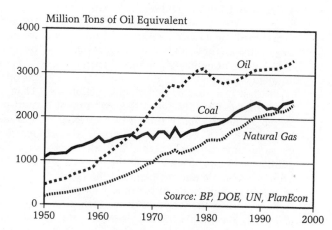

Figure 2: Fossil Fuel Use by Fuel Type, 1950–96

Nuclear Power Inches Up   Nicholas Lenssen

Between 1995 and 1996, total installed nuclear generating capacity increased by less than 1 percent, bringing the total to a new high of 343,586 megawatts.[1] (See Figure 1.) Since 1990, global nuclear capacity has risen less than 5 percent, in distinct contrast to a growth rate of 9 percent a year in the 1980s.[2]

Altogether, 436 reactors were listed as grid-connected at the end of 1996.[3] The small gain in global nuclear capacity came from the completion of one reactor each in France, Japan, Romania, and the United States.[4] These openings were partly offset by the closure of two units (in Ukraine and the United States), bringing to 86 the number of reactors that have been retired after an average service life of less than 17 years.[5] (See Figure 2.)

In 1996, construction started on nine new reactors—the most since 1985—as four new units were started in South Korea, two each in China and Taiwan, and one in Japan.[6] (See Figure 3.) Worldwide, some 37 reactors (with a combined capacity of just under 30,000 megawatts) are now under active construction—representing just 9 percent of current installed capacity.[7]

The last U.S. reactor under construction was completed in 1996 when the government-owned Tennessee Valley Authority connected Watts Bar 1 to the electric power grid.[8] At the same time, privately owned U.S. electric utilities were busy trying to cut their losses from uneconomic nuclear plants in the face of growing competition in the power industry.

Elsewhere in the world, Ukraine took a major step toward closing the Chernobyl station with the permanent closure of Unit 1.[9] In Russia, a shortage of funds has left nuclear workers at operating plants unpaid for months, as well as stalled construction on new plants.[10] In December 1996, in a referendum, citizens of the Russian Kostroma region rejected plans to build a nuclear station.[11]

In Western Europe, France is the only country still building nuclear plants, with three units in the pipeline.[12] Meanwhile, the Dutch Board of Electricity Producers announced it would permanently close one of two reactors in the country in early 1997.[13]

In mid-1996, the British government sold some of the country's nuclear plants to investors, though it raised only $2.2 billion in the process—less than it initially expected.[14] This did not even equal the $4.53-billion price tag for the single reactor, Sizewell B, that came on-line in 1995.[15]

Asia is the only region where nuclear power is likely to expand significantly in the next decade, though no longer in Japan, which currently has 70 percent of regional capacity.[16] By early 1997, Japan only had four reactors under construction (including the Monju fast breeder reactor, which had a major accident during start-up testing in 1995).[17]

Political support for nuclear power in Japan has continued to erode since the Monju accident. In August 1996, in the country's first public referendum on nuclear power, the residents of the town of Maki voted against allowing a plant to be built.[18] And in September, the governor of Fukui Prefecture—site of the Monju plant—rejected new government subsidies to site additional reactors there.[19]

South Korea currently leads the world in construction, with 9 reactors being built and 11 in operation.[20] Local opposition stalled the start of construction on two reactors for nearly all of 1996, however, and the government is still unable to site radioactive waste facilities due to public opposition.[21] China may pass South Korea as the world's leading builder: official plans call for an increase from the current 2,100 megawatts of capacity to 20,000 megawatts by 2010.[22] Work started on two new reactors in 1996, with plans to add four more to the pipeline in 1997.[23]

Finally, in Taiwan, the government forced through a divided parliament the construction budget for two reactors first proposed in 1982.[24] In May, opposition lawmakers cancelled funding, but the government restored it in October when the opposition failed to muster the two-thirds majority needed to block the project.[25] And Turkey may soon enter the nuclear club, as the government put out a tender in late 1996 to test the waters for that country's first order.[26]

WORLD NET INSTALLED
ELECTRICAL GENERATING CAPACITY
OF NUCLEAR POWER PLANTS,
1960–96

YEAR	CAPACITY (gigawatts)
1960	1
1961	1
1962	2
1963	2
1964	3
1965	5
1966	6
1967	8
1968	9
1969	13
1970	16
1971	24
1972	32
1973	45
1974	61
1975	71
1976	85
1977	99
1978	114
1979	121
1980	135
1981	155
1982	170
1983	189
1984	219
1985	250
1986	276
1987	297
1988	310
1989	320
1990	328
1991	325
1992	327
1993	336
1994	338
1995	340
1996 (prel)	344

SOURCE: Worldwatch Institute database,
compiled from the IAEA and press reports.

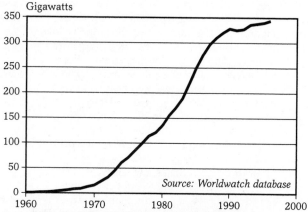

**Figure 1: World Electrical Generating Capacity of
Nuclear Power Plants, 1960–96**

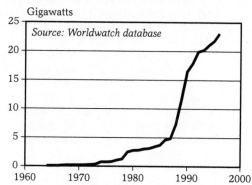

**Figure 2: Cumulative Generating Capacity
of Closed Nuclear Power Plants, 1964–96**

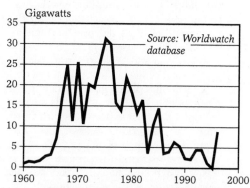

**Figure 3: World Nuclear Reactor
Construction Starts, 1960–96**

Geothermal Power Rises   Seth Dunn

Geothermal electricity generating capacity grew by 375 megawatts—around 5.5 percent—to 7,173 megawatts in 1996.[1] (See Figure 1.) (The capacity is substantially lower than reported in *Vital Signs 1993*, which relied on U.N. data for geothermal energy that actually covered all renewable sources.) The tapping of geothermal energy—heat from the center of the Earth—provides electricity in 22 countries, most of them in the developing world.[2] But these estimated 250 power plants generate less than 1 percent of total world power.[3]

In the United States, currently the leading user, geothermal energy supplies enough power for nearly a million people.[4] An estimated 2,800 megawatts are currently installed in California, Hawaii, Nevada and Utah.[5] But only 30 megawatts have been added since 1990, with none installed in 1996, as insufficient heat from fields has stalled several projects.[6] (See Figure 2.)

The Philippines, the world's second largest user (approximately 1,200 megawatts), has added more than 300 megawatts since 1990 and is projected to tack on another 800 by 2000, in part by rehabilitating existing plants.[7] Mexico, the world's third leading user (some 750 megawatts), is set to add another 200 megawatts by 2000.[8]

Indonesia has more than doubled its geothermal capacity since 1990 and may triple installations by 2000.[9] Since 1990, Japan has nearly doubled geothermal capacity.[10] China is believed to have abundant geothermal resources: experts project that by 2000 the nation will nearly quadruple its capacity from a modest 30 megawatts.[11]

Nicaragua and El Salvador are planning geothermal projects; Costa Rica, which had no geothermal plants in 1990, now has 55 megawatts in use.[12] Argentina is currently the only geothermal power user in South America.[13] Several Caribbean islands have begun constructing plants as well.[14]

Central Asia and Africa have some, but less, geothermal activity. Turkey has just 21 megawatts from this source, though a sixfold increase is projected by 2000.[15] By that time,

Russia, which generates only 11 megawatts at present, is expected to make a tenfold jump.[16] Kenya plans to double its 45 megawatts by 2000, while Zambia has a small 200-kilowatt plant installed.[17]

In Europe, growth in geothermal power is small but steady. Italy has more than 600 megawatts, 90 of them added during the 1990s.[18] Germany, France, and Portugal have little capacity; Greece shut down its only pilot plant due to opposition from environmental groups during the 1980s.[19]

Geothermal energy raises several problems despite its relatively benign characteristics. The hydrogen sulfide emitted by geothermal power plants has caused health problems, while cultural concerns have blocked development elsewhere—near, for example, the U.S. Yellowstone National Park, the world's largest geyser collection.[20] Large-scale geothermal power plants also require substantial upfront investments.[21]

Simpler and cleaner ways to use and tap geothermal heat are being tested. One method generates electricity from lower temperatures by transferring heat from geothermal fluid to a fluid with a lower boiling point, such as pentane. Another technique, called hot dry rock, injects water into hot fissures and then retrieves it; these are more common than the natural reservoirs of water and steam that geothermal plants typically exploit.[22]

Geothermal energy also has important direct applications, though potentially at higher costs. Such uses, especially heat pumps to insulate buildings, are common on all continents.[23] In Iceland, geothermal energy provides heat for 85 percent of the population, mostly through space heating.[24]

Some 31 plants are under discussion in 11 countries and exploration is rising, particularly in New Zealand and other Pacific islands.[25] But as geothermal energy use increases, conflicts are likely to erupt as well. In the Philippines, for example, the government plans to continue building a 120-megawatt plant on Mount Apo despite the opposition of the indigenous Ladi, to whom the mountain is sacred.[26]

WORLD GEOTHERMAL POWER,
1950–96

YEAR	WORLD (megawatts)
1950	200
1955	262
1960	374
1965	556
1966	579
1967	661
1968	677
1969	704
1970	711
1971	817
1972	920
1973	1,123
1974	1,124
1975	1,287
1976	1,320
1977	1,380
1978	1,451
1979	1,946
1980	2,471
1981	2,531
1982	2,898
1983	3,451
1984	3,912
1985	4,414
1986	4,667
1987	5,007
1988	5,070
1989	5,154
1990	5,832
1991	6,000
1992	6,275
1993	5,900
1994	6,170
1995	6,798
1996 (prel)	7,173

SOURCES: Mary H. Dickson and Mario Fanelli, "Geothermal Energy Worldwide," *World Directory of Renewable Energy Suppliers and Services 1995*; Mary Dickson, International Institute for Geothermal Research, letter to author, 3 February 1997.

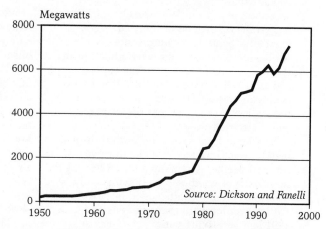

Source: Dickson and Fanelli

Figure 1: World Geothermal Power Installed, 1950–96

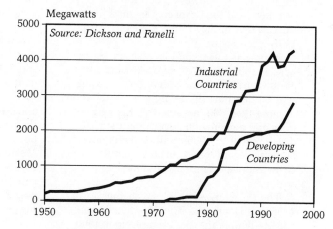

Source: Dickson and Fanelli

Figure 2: World Geothermal Power Installed in Industrial and Developing Countries, 1950–96

Wind Power Growth Continues Christopher Flavin

Global wind power generating capacity surged 26 percent in 1996, reaching 6,070 megawatts by the end of the year.[1] (See Figure 1.) Some 1,290 megawatts of new wind generating capacity were added during the year, virtually level with the capacity added in 1995.[2] (See Figure 2.) Wind power is now the fastest growing energy source, though it still produces less than 1 percent of world electricity.[3]

The largest market for new wind turbines in 1996 was Germany—for the third year in a row. Altogether, Germany added 426 megawatts of wind power—down slightly from the 500 megawatts added in 1995.[4] Battles over local siting regulations and a 1991 law permitting wind generators to receive 17 pfennigs (10¢) per kilowatt-hour for the electricity they generate slowed the German wind boom early in 1996.[5] By the end of the year, however, development had resumed, and total capacity seemed likely to surpass that in the United States by April 1997. (See Figure 3.)

India was the second largest market for wind turbines in 1996; some 244 megawatts of new turbines were added, less than the year before, but still taking total installations to more than 800 megawatts.[6] Rising interest rates and a change in government slowed development in 1996. Still, India is now home to several wind turbine manufacturers. Growth prospects for 1997 and beyond are strong.[7]

The third largest wind power market in 1996 was Denmark. The 200 megawatts added—double the 1995 figure—took total installations to 835 megawatts.[8] Wind turbines are now found virtually throughout rural areas of this small country.[9] The government estimates that wind power is one of the least expensive ways to reduce Danish carbon emissions.[10] Denmark's pioneering wind power companies also dominate international markets, thanks in part to government export assistance.

Spain's wind market surged to fourth largest in 1996, adding 116 megawatts and nearly doubling its capacity during the course

of the year.[11] A recent Spanish electricity law provides a generous purchase price for renewable power projects, which has spurred a number of Spanish companies to enter the business.[12]

The British wind power market, which had been expected to take off in 1996, had another lackluster year. Only 73 megawatts were added, according to preliminary estimates, despite the fact that the country has Western Europe's greatest wind power potential.[13] The industry was held back by local siting problems and unattractive purchase prices offered through its national bidding scheme. The Dutch wind industry also had a slow 1996, with just 48 megawatts added.[14]

China's wind industry began to show life in 1996.[15] Although the country added just 35 megawatts, foreign companies reported another 70 megawatts of new orders at the end of the year, suggesting that 1997 may be a big year for China.[16] With power demand far outstripping supply, a vibrant and rapidly expanding private sector, and some of the world's most abundant wind resources, China is the odds-on favorite to become the dominant wind power market in the next decade.

The western hemisphere remained a virtual desert for wind energy in 1996, with only the odd turbine or two added in such potential wind "powers" as Brazil, Canada, and Mexico. In the United States—which still barely leads the world in total wind power capacity, with 1,596 megawatts—a meager 12 megawatts were added in 1996. At least 110 megawatts of old turbines have been removed in California since 1994, and the country's total wind capacity has been on a slow downward path since 1991.[17]

Kenetech, the leading U.S. wind company, declared bankruptcy in 1996, but in January 1997, Enron—a multibillion-dollar gas and electricity supplier—purchased the Zond Corporation, the largest remaining wind firm, and brought a potential infusion of capital to the U.S. wind industry.[18]

WORLD WIND ENERGY GENERATING
CAPACITY, 1980–96

YEAR	CAPACITY (megawatts)
1980	10
1981	25
1982	90
1983	210
1984	600
1985	1,020
1986	1,270
1987	1,450
1988	1,580
1989	1,730
1990	1,930
1991	2,170
1992	2,510
1993	2,990
1994	3,680
1995	4,821
1996 (prel)	6,070

NET ANNUAL ADDITIONS TO WORLD
WIND GENERATING CAPACITY,
1980–96

YEAR	CAPACITY (megawatts)
1980	5
1981	15
1982	65
1983	120
1984	390
1985	420
1986	250
1987	180
1988	130
1989	150
1990	200
1991	240
1992	340
1993	480
1994	720
1995	1,294
1996 (prel)	1,290

SOURCES: Birger Madsen, BTM Consult, Denmark,
17 January 1997; Paul Gipe and Associates,
Tehachapi, CA, discussion with author, 19 February
1996; Knud Rehfeldt, Deutsches Windenergie-
Institut, letter to author, 13 January 1997.

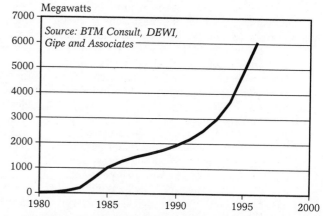

Figure 1: World Wind Energy Generating Capacity,
1980–96

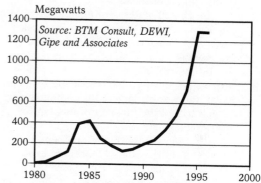

Figure 2: Net Annual Additions to World
Wind Energy Generating Capacity, 1980–96

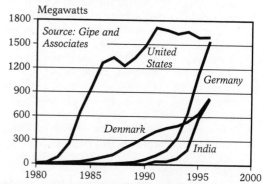

Figure 3: Wind Generating Capacity in the United
States, Germany, Denmark, and India, 1980–96

Solar Cell Shipments Keep Rising · Molly O'Meara

Global shipments of photovoltaic (PV) cells—the semiconductors that turn sunlight into electricity—reached 89.6 megawatts in 1996, up from 78.6 megawatts in 1995.[1] (See Figure 1.) This was the third straight year of double-digit growth since a slowdown in the early 1990s, although the 15.5-percent increase in 1996 was less than anticipated by industry analysts.[2]

Cumulative shipment of PV cells to date grew in 1996 to nearly 700 megawatts.[3] (See Figure 2.) As in 1995, supply kept pace with demand in 1996, so wholesale factory prices for PV cells remained around $4 per watt.[4] (See Figure 3.) Most PV cells in 1996 were made of silicon in single crystal or polycrystalline form; in 1997, several manufacturers are increasing their capacity to make less costly thin-film silicon.[5]

The United States continued to dominate the PV market, with shipments increasing from 34.75 megawatts in 1995 to 39.85 megawatts.[6] These sales are expected to double by the end of 1998.[7] One of the major expansions during 1996 was made by Siemens Solar, the world's largest PV manufacturer, which tripled the capacity of its plant in Washington state.[8]

At present, more than 70 percent of the PVs made in the United States are exported.[9] The top destinations in 1996 were Germany, Japan, Mexico, Netherlands, and the United Kingdom.[10] To encourage the development of its domestic market, the U.S. government is launching a "Buildings for the 21st Century" campaign in 1997, aimed at installing solar cells in 1 million buildings by 2010.[11]

Japanese shipments jumped from 16.4 megawatts in 1995 to 21.2 megawatts in 1996, spurred by domestic demand for solar homes.[12] The Japanese government paid half the cost of 3,000 solar home systems for consumers in 1995 and 1996.[13] Sharp Corporation, which makes PV arrays, anticipates that PV systems will be installed in 60,000—10 percent—of all new houses in Japan by 2000.[14]

In 1996, European output of PVs declined slightly from 20.1 megawatts in 1995 to 18.8 megawatts.[15] Large companies such as ASE GmbH and Siemens have been shifting production to their U.S. plants, and governments have been reducing their incentive programs.[16]

A 1996 survey of manufacturers found that the PV market in 1995 broke down as follows: 45 percent of PV cells were used to electrify homes, villages, and water-pumping systems; 36 percent went to communications and other remote industrial applications; 14 percent were used for electrical generation linked to grid systems; and 5 percent were used to power calculators, watches, and other small products.[17] Scientists are researching new applications—for instance, some of the communications functions of conventional satellites may be replaced with PV-powered, unmanned, ultralight aircraft.[18]

The greatest potential for PV expansion is in developing countries, where roughly 2 billion people lack electricity, and more than 400,000 houses already use PV systems.[19] These systems typically include a small, 25–55 watt array of PV cells, a rechargeable battery to store electricity for use at night, one or more lights, and an outlet for appliances.[20]

But household adoption of PV electricity has been hampered by potential consumers' lack of access to credit from banks or PV distributors.[21] New lending schemes in such countries as Bangladesh, China, India, and Viet Nam indicate that market obstacles are beginning to be overcome.[22] In India, for example, where more than 500 million rural dwellers have no electricity, the Syndicate Bank teamed up in 1996 with the Solar Electric Light Company to offer customers both credit and service for PV home systems.[23]

To support the PV industry, the World Bank approved a PV Market Transformation Initiative in 1997 to provide $30 million to companies or consortia that present proposals for accelerating PV development in India, Kenya, and Morocco.[24]

WORLD PHOTOVOLTAIC SHIPMENTS, 1971–96

YEAR	SHIPMENTS (megawatts)
1971	0.1
1975	1.8
1976	2.0
1977	2.2
1978	2.5
1979	4.0
1980	6.5
1981	7.8
1982	9.1
1983	17.2
1984	21.5
1985	22.8
1986	26.0
1987	29.2
1988	33.8
1989	40.2
1990	46.5
1991	55.4
1992	57.9
1993	60.1
1994	69.4
1995	78.6
1996 (prel)	89.6

SOURCE: Paul Maycock, *PV News.*

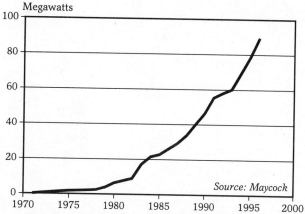

Figure 1: World Photovoltaic Shipments, 1971–96

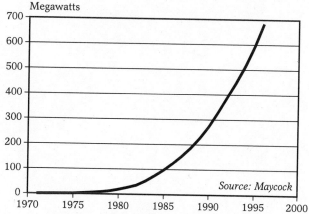

Figure 2: World Photovoltaic Shipments, Cumulative, 1971–96

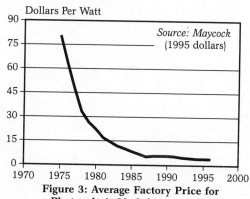

Figure 3: Average Factory Price for Photovoltaic Modules, 1975–96

Atmospheric Trends

Carbon Emissions Set New Record Seth Dunn

Worldwide carbon emissions from the burning of fossil fuels climbed to 6.25 billion tons in 1996, reaching a new high for the second year in a row.[1] (See Figure 1.) This increase of just over 2.8 percent, the largest since 1988, reflects a steady rise in emissions in the developing world, a slowing of growth in industrial countries, and a flattening of steep declines in the former Eastern bloc. (See Figure 2.) Global carbon emissions have now nearly quadrupled since 1950.[2]

Carbon output in the industrial world continued to grow, albeit at a slower rate than in the 1960s and 1970s. The United States—the world leader, with nearly 23 percent of the total—increased carbon output more than 8 percent between 1990 and 1996.[3] Japan, with 5 percent of the world total, saw emissions rise more than 8 percent from 1990 to 1995—though its per capita levels are still less than half those in the United States.[4]

In regions in transition, emissions are levelling off after a dramatic drop in the early 1990s. In reunified Germany, emissions fell more than 10 percent between 1990 and 1995.[5] In the former Eastern bloc, emissions dropped 10 percent in Poland and 28 percent in Russia.[6]

Several rapidly industrializing countries are experiencing steep emissions growth. Brazil, India, and Indonesia increased emissions 20, 28, and 40 percent respectively between 1990 and 1995.[7] But their per capita output is still less than one tenth that in the United States.[8]

The most significant rise is taking place in China, where double-digit, coal-driven economic growth has boosted emissions more than 27 percent since 1990.[9] China now contributes 14 percent of the global carbon output, the second highest. Its emissions per capita, though, remain below one seventh those in the United States.[10]

When released through combustion, carbon reacts with oxygen to form carbon dioxide (CO_2)—the most common of the greenhouse gases that trap heat in the atmosphere, causing surface temperatures to rise. Industrialization has pushed emissions past the rate of uptake by forests and oceans, which yearly absorb 3–4 billion tons of carbon.[11] Since the mid-nineteenth century, atmospheric CO_2 levels have risen 29 percent—from 280 to 362 parts per million, their highest point in the last 150,000 years.[12]

Carbon emissions are contributing to changes in the Earth's climate. In 1995 the Intergovernmental Panel on Climate Change, a U.N. panel of 2,500 scientists, concluded that "the balance of evidence suggests a discernible human influence on global climate."[13] It projected that a doubling of greenhouse gas concentrations would increase global temperature by 1–3.5 degrees Celsius by the year 2100, which could cause widespread economic, social, and environmental dislocations.[14] But the panel also noted that significant reductions in carbon emissions are both technically and economically feasible.[15]

Some 160 countries have signed the U.N. Framework Convention on Climate Change, aimed at stabilizing greenhouse gas concentrations at levels that will avoid dangerous climate change. It commits industrial countries to try to return carbon emissions to 1990 levels by 2000.[16] Most are likely to miss the target, however: the United States is projected to exceed the mark by 11 percent, and the European Union and Japan, by 6 percent.[17]

Yet there are some signs of progress in slowing emissions growth: Germany, the United Kingdom, and France may meet the 2000 goal.[18] Ambitious plans exist in smaller countries like the Netherlands—whose per capita emissions are near those of Japan—and in developing countries, which are highly vulnerable to the impacts of climate change.[19]

Nevertheless, without additional policy initiatives, the upward trend will continue: the International Energy Agency projects annual emissions could reach nearly 9 billion tons by 2010—49 percent above 1990 levels.[20] Negotiations are under way for a legally binding agreement—to be considered in Kyoto, Japan, in December 1997—to commit industrial countries to reducing emissions beyond the year 2000.[21]

WORLD CARBON EMISSIONS FROM
FOSSIL FUEL BURNING, 1950–96

YEAR	EMISSIONS (mill. tons of carbon)
1950	1,620
1955	2,020
1960	2,543
1965	3,095
1966	3,251
1967	3,355
1968	3,526
1969	3,735
1970	4,006
1971	4,151
1972	4,314
1973	4,546
1974	4,553
1975	4,527
1976	4,786
1977	4,920
1978	4,960
1979	5,239
1980	5,170
1981	4,998
1982	4,959
1983	4,945
1984	5,114
1985	5,286
1986	5,472
1987	5,593
1988	5,809
1989	5,914
1990	5,943
1991	6,010
1992	5,926
1993 (est)	5,919
1994 (est)	5,989
1995 (est)	6,080
1996 (prel)	6,251

SOURCES: Marland, Andres, and Boden, electronic
database, Oak Ridge National Laboratory, 1995;
1993–96, Worldwatch estimates based on Marland,
Andres, and Boden, on OECD, and on BP.

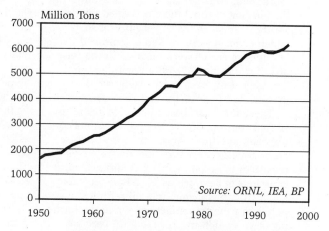

**Figure 1: World Carbon Emissions from
Fossil Fuel Burning, 1950–96**

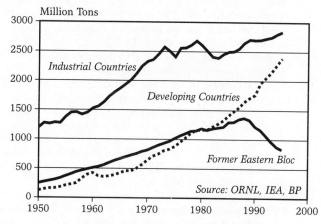

**Figure 2: Carbon Emissions from
Fossil Fuel Burning, by Economic Region, 1950–96**

World emissions of sulfur and nitrogen from the burning of fossil fuels remained virtually steady in 1994, the most recent year for which estimates are available. Some 70.7 million tons of sulfur were released in the form of sulfur dioxide (SO_2), along with 28.2 million tons of nitrogen in nitrogen oxides (NO_x).[1] (See Figures 1 and 2.)

These numbers, based on estimates of fossil fuel consumption rates and emissions controls, are unofficial approximations because no international organization tracks the global release of these pollutants.[2] Additional sulfur emissions of as much as 15–20 percent may result from metal smelting, oil refining, and pulp and paper manufacture, while the burning of vegetation may produce 30 percent more nitrogen emissions.[3]

Although sulfur dioxide and nitrogen oxides also arise from natural processes—SO_2 from volcanoes and sea spray; NO_x from soil bacteria, algae, and lightning—the increase in fossil fuel use since the Industrial Revolution has led to harmful levels of these gases.

Both SO_2 and NO_x cause acid rain, which damages forests, aquatic life, and crops. The acids formed by sulfur dioxide also corrode the stone and metal of buildings.[4] Nitrogen oxides, released in great quantities from cars that lack catalytic converters, are precursors of urban smog.

Relatively steady global sulfur and nitrogen emissions since the late 1980s are the result of increased pollution from heavier coal use in developing countries being countered by falling emissions from industrial nations, which have seen gains in energy efficiency, cleaner technologies, and cleaner fuels, such as natural gas.[5]

In the United States and Western Europe, the 1979 Convention on Long Range Transboundary Air Pollution helped shape domestic regulations that encourage cleaner industrial practices.[6] The U.S. Clean Air Act regulates sulfur dioxide emissions through tradable pollution permits. Corporations have found it easier than expected to comply, by switching to low-sulfur fuels, so the cost of these permits has plummeted: at the 1996 auction, permits to emit one ton of sulfur dioxide sold at around $60, down from $130 in 1995.[7]

Nitrogen pollution is more difficult to abate, because much of it comes from vehicles. While U.S. emissions of SO_2 dropped by 37 percent from 1986 to 1995, those of NO_x fell by just 14 percent.[8]

In the European Union, SO_2 emissions fell by 27 percent between 1990 and 1994, whereas NO_x emissions were down by only 10 percent.[9] Of the 12 European countries that pledged in 1988 to reduce their NO_x emissions by some 30 percent by 1998, only half appear to be achieving that goal.[10]

The most visible effects of sulfur and nitrogen emissions are found in the cities of the developing world. China, which has reported at least 3 million deaths from urban air pollution since 1994, pledged in 1996 to strengthen curbs on sulfur dioxide and to force more than 70,000 industrial polluters to close.[11] In India, the Supreme Court ordered nearly 300 coal-based industries surrounding the Taj Mahal to shut down by the end of 1997 because the building has been damaged by sulfur dioxide.[12] If industrializing countries increase fossil fuel use without tightening pollution controls, global sulfur and nitrogen emissions may again push upwards.

Nitrogen, the atmosphere's most abundant element, is an essential plant nutrient, but only a small amount is converted by natural processes into forms that plants can use. Fossil fuel combustion and fertilizer production may be doubling the amount of nitrogen in nutrient form, with myriad implications for life on earth.[13] In lakes and oceans, for instance, excess nitrogen spurs the growth of algae that choke off the growth of other species.[14] Some have speculated that increased fertilization of plants, which take up carbon dioxide, will offset global warming.[15] A 12-year study found that this theory did not hold up in grasslands, where added nitrogen favored grasses that were less effective at sequestering carbon dioxide than the species they replaced.[16]

WORLD SULFUR AND NITROGEN
EMISSIONS FROM FOSSIL FUEL
BURNING, 1950–94

YEAR	SULFUR (mill. tons)	NITROGEN (mill. tons)
1950	30.1	6.8
1960	46.2	11.8
1970	57.0	18.1
1971	56.9	18.6
1972	58.2	19.5
1973	60.9	20.6
1974	60.9	20.8
1975	56.4	19.9
1976	58.6	21.0
1977	60.1	20.8
1978	61.0	22.3
1979	62.6	22.4
1980	62.9	22.3
1981	61.9	22.1
1982	62.1	22.2
1983	63.0	22.5
1984	64.5	23.3
1985	64.2	23.4
1986	65.2	23.6
1987	66.5	24.3
1988	71.2	27.3
1989	71.9	27.7
1990	70.0	27.7
1991	71.1	28.0
1992	70.4	27.8
1993	70.4	28.0
1994	70.7	28.2

SOURCES: J. Dignon, letter to author, 22 January 1997;
Hameed and Dignon, *Journal of the Air & Waste
Management Association*, February 1992; Dignon and
Hameed, *JPCA*, February 1989.

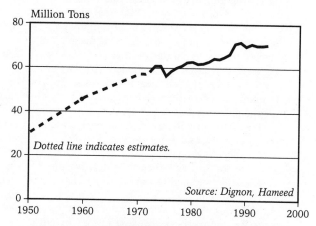

**Figure 1: World Sulfur Emissions from Fossil Fuel
Burning, 1950–94**

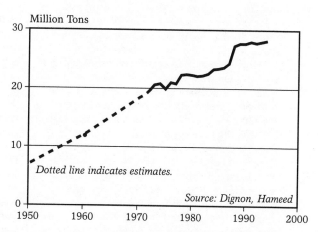

**Figure 2: World Nitrogen Emissions from Fossil Fuel
Burning, 1950–94**

Global Temperature Down Slightly   Seth Dunn

The temperature of the atmosphere at the Earth's surface averaged 15.32 degrees Celsius in 1996, according to preliminary figures, placing it among the five warmest years since data collection began in 1866.[1] (See Figure 1.) Though this is a slight drop from the 1995 high of 15.40 degrees Celsius, global temperatures have increased nearly half a degree since 1950.[2]

The 1990s are already the warmest decade on record—averaging 0.1 degrees Celsius above the 1980s—according to the Goddard Institute for Space Studies at NASA, which collects the land and ocean surface-temperature measurements.[3]

The warmth of the current decade is particularly remarkable because it has occurred in conjunction with several short-term natural and humanmade cooling effects. These include the century's largest volcanic event, the 1991 eruption of Mount Pinatubo; the solar energy cycle, which has been at a minimum during the 1990s; and atmospheric depletion of ozone, now at record levels.[4]

More recent cooling influences also affected 1996 temperatures. The presence of La Niña, an upwelling of unusually cool waters in the equatorial Pacific Ocean, had a role in the temperature drop. Also partly responsible was the reversal of the North Atlantic Oscillation, a 30-year trend of cooling in Greenland and warming in North America and Europe, leading to record precipitation and extreme cold events in the two latter regions during 1996.[5]

According to data from the Hadley Centre and the University of East Anglia, 1996 continued an underlying upward trend begun in the mid-1970s, with some regions warming quickly.[6] Summer temperatures in northern Siberia are warmer than they have been in a millennium, forcing boreal forests northward.[7] Antarctica has warmed at more than twice the average global rate during the last 50 years, causing five of the continent's ice shelves to disintegrate.[8]

Rising atmospheric temperatures interact dynamically with ocean processes. Geological records and computer models reveal that the ocean's heat-carrying conveyor belt shifts suddenly in response to temperature changes—leading to abrupt climate changes such as dramatic cooling in northern Europe—which may reduce the ocean's ability to absorb carbon.[9] Warming also causes oceans to lose nitrate, slowing the growth of carbon-assimilating phytoplankton.[10]

Many aquatic, marine, and terrestrial ecosystems are highly sensitive to small temperature increases: freshwater fish, coral reefs, and boreal forests are particularly at risk.[11] Warming, moreover, behaves synergistically with ozone depletion and acidification to compound ecological stresses.[12] And it can feed on itself in certain instances: the loss of boreal forest and warming of tundra could release large amounts of carbon dioxide as well as methane, another potent greenhouse gas.[13]

Feedbacks from the ocean and biosphere as the atmosphere warms are examples of climate's tendency to behave unexpectedly when rapidly forced to change.[14] Such "surprises," which have occurred in the past but are difficult to predict, could increase the rate of warming—which is already expected to be the fastest seen in 10,000 years.[15] This climate instability poses serious and widespread risks to human health, according to a 1996 report prepared for the World Health Organization, the United Nations Environment Programme, and the World Meteorological Organization.[16]

Evidence of the human "fingerprint" in climate change continues to strengthen with improved understanding of sulfates and other influences on the atmosphere's temperature.[17] (See Figure 2.) A team led by Goddard's James Hansen has clarified the relationship between these influences and observed global temperature changes, and suggests there could be a return to the warming trend as the La Niña effect fades.[18] Hansen believes there is a "high likelihood" that another temperature record will be set before the end of the century.[19]

GLOBAL AVERAGE TEMPERATURE, 1950–96

YEAR	TEMPERATURE (degrees Celsius)
1950	14.86
1955	14.92
1960	14.98
1965	14.88
1966	14.95
1967	14.99
1968	14.93
1969	15.05
1970	15.02
1971	14.93
1972	15.00
1973	15.11
1974	14.92
1975	14.92
1976	14.82
1977	15.11
1978	15.05
1979	15.09
1980	15.18
1981	15.29
1982	15.08
1983	15.24
1984	15.11
1985	15.09
1986	15.16
1987	15.27
1988	15.28
1989	15.22
1990	15.39
1991	15.36
1992	15.11
1993	15.14
1994	15.23
1995	15.40
1996 (prel)	15.32

SOURCE: Goddard Institute for Space Studies, New York, 14 January 1997.

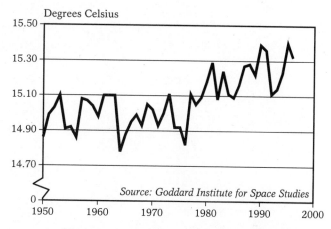

Figure 1: **Average Temperature at the Earth's Surface, 1950–96**

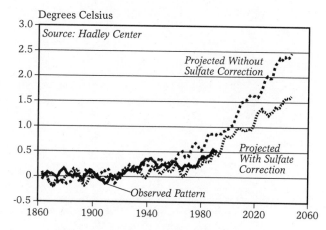

Figure 2: **Models of Global Warming Compared With Observations, 1863–2047**

Economic Trends

World Economy Expands Faster

Lester R. Brown

World output of goods and services in 1996 expanded by 3.8 percent, up modestly from the 3.5-percent growth in 1995.[1] The gross world product (GWP) climbed from $26.9 trillion to $28.0 trillion (1995 dollars), and GWP per person increased from $4,733 to $4,846, a gain of 2.4 percent.[2] (See Figures 1 and 2.)

Two important points stand out in the data for 1996. Developing countries grew at 6.3 percent in 1996, nearly three times as fast as industrial nations.[3] And the transition economies—in Eastern Europe and the former Soviet Union—managed a slight expansion in 1996, reversing five consecutive years of decline when they were adjusting to the breakup of the Soviet Union and the shift to a market economy.[4]

Industrial countries grew by 2.3 percent in 1996 versus 2.1 percent in the preceding year.[5] The trend among countries was mixed. Germany, France, and Italy all saw their growth rates cut roughly in half, but the United Kingdom maintained its steady expansion.[6] The U.S. economy grew by 2.4 percent in 1996, up from 2.0 percent in 1995.[7] The big change came in Japan, where the economy expanded by 3.5 percent in 1996 after four years in which growth was negligible or nonexistent.[8]

The transition economies of Eastern Europe and the former Soviet Union had an overall growth rate of 0.4 percent in 1996, the first rise in the regional economy in five years.[9] Some of these countries have turned the corner with economic reforms and are expanding, such as Albania, the Czech Republic, Poland, Armenia, Mongolia, Latvia, Lithuania, and Estonia; others, such as Russia and the Ukraine, continued their decline in 1996.[10]

The 6.3-percent growth in the economies of developing countries in 1996 marked the fifth consecutive year of growth at 6 percent or more.[11] Asia (excluding Japan) grew at 8 percent in 1996, its fifth year of growth at that level or better.[12] Economic growth in China dropped to 9 percent in 1996, ending four years of double-digit increases in which the economy grew by more than half.[13]

Viet Nam, at 9.5 percent in 1996, has exceeded 8 percent each year since 1992.[14] Other countries with rapid growth in 1996 included Malaysia at 8.8 percent, Thailand at 8.3 percent, and Indonesia at 7.8 percent.[15] The Indian subcontinent maintained steady growth, with the three major economies—India, Bangladesh, and Pakistan—collectively increasing by 6 percent.[16]

Latin America came in at roughly 3 percent in 1996 as the region recovered from the Mexican peso devaluation crisis that had dropped its economic growth to 0.9 percent the preceding year.[17] The field was led by a 7.4-percent expansion in Chile, the region's strongest economy.[18] Among the larger economies, Brazil and Argentina both expanded at 2.5 percent.[19] Mexico grew by more than 4 percent after shrinking by nearly 7 percent in 1995 in the wake of the peso crisis.[20]

Africa continued to strengthen its economic performance in 1996, achieving an estimated growth of 5 percent.[21] The region was led by Morocco at 9.2 percent and Tunisia at 7.5 percent, both recovering from a severe drought the previous year.[22] The East African leaders were Uganda at 6 percent and Tanzania and Kenya, both at 5 percent.[23]

The pacesetter in West Africa was Côte d'Ivoire at 6.5 percent, achieved in part by developing its oil resources.[24] It was followed by Cameroon and Ghana, both expanding at 5 percent.[25] South Africa, gaining momentum after political reforms, grew 4 percent in 1996.[26]

Economic reforms designed to reduce fiscal deficits and check inflation, combined with other economic reforms needed to attract more foreign investment, helped set the stage for more rapid economic expansion in many of these developing countries.

In preliminary estimates for 1997, the International Monetary Fund is projecting world economic growth at 4.1 percent.[27] This gain will be fed by acceleration in Germany, France, and Italy from roughly 1.2 percent to 2.4-percent growth in 1997, somewhat faster growth in Latin America, and a big jump in the transition economies as they go from 0.4 percent to 4 percent.[28]

Gross World Product, 1950–96

YEAR	TOTAL (trill. 1995 dollars)	PER CAPITA (1995 dollars)
1950	4.9	1,925
1955	6.3	2,283
1960	7.9	2,599
1965	10.2	3,059
1966	10.7	3,147
1967	11.1	3,196
1968	11.8	3,314
1969	12.6	3,460
1970	13.1	3,529
1971	13.6	3,594
1972	14.2	3,690
1973	15.2	3,848
1974	15.3	3,808
1975	15.4	3,770
1976	16.2	3,891
1977	16.8	3,977
1978	17.5	4,061
1979	18.1	4,138
1980	18.3	4,096
1981	18.5	4,084
1982	18.6	4,041
1983	19.2	4,082
1984	19.9	4,177
1985	20.7	4,268
1986	21.2	4,302
1987	22.0	4,383
1988	23.1	4,511
1989	23.8	4,587
1990	24.3	4,610
1991	24.2	4,513
1992	24.6	4,520
1993	25.2	4,557
1994	26.0	4,631
1995	26.9	4,733
1996 (prel)	28.0	4,846

SOURCES: GWP data for 1950 and 1955 from Herbert R. Block, *The Planetary Product in 1980: A Creative Pause?* (Washington, DC: U.S. Department of State, 1981); World Bank and International Monetary Fund tables.

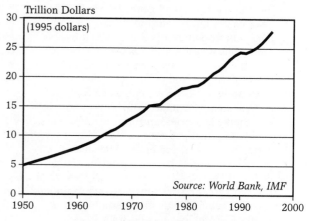

Figure 1: Gross World Product, 1950–96

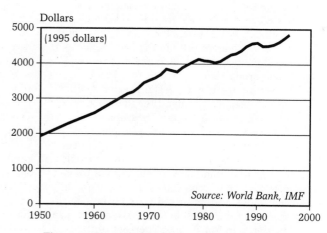

Figure 2: Gross World Product Per Person, 1950–96

Roundwood Production Rises Again
Cheri Sugal

After a slight decline in 1991, roundwood production has continued a steady rise that has been nearly unbroken since 1950. In 1995, total roundwood production reached 3.5 billion cubic meters, an increase of 0.6 percent over the previous year and more than twice the 1950 production level.[1] (See Figure 1.)

Roundwood is used to produce fuelwood and charcoal, a primary source of energy in developing countries, and industrial roundwood—logs that get cut into boards for construction or wood products. The share used for fuelwood and charcoal rose from 50 percent in 1980 to 56 percent in 1995.[2] Some of this increase, however, may be due to underrepresentation of industrial roundwood as a result of illegal logging and trade.[3]

The relative contribution of industrial production and fuelwood extraction differs greatly between industrial and developing countries. In the former, industrial production accounted for 85 percent of total roundwood production in 1995.[4] (See Figure 2.) In developing countries, which are mostly in the tropics, fuelwood extraction took 80 percent of roundwood production that year.[5] (See Figure 3.)

Within the tropics, certain regional differences exist. The demand for fuelwood is a problem primarily in the dry tropics.[6] In Mali and Burkino Faso, for example, fuelwood extraction accounts for about 95 percent of all tree cutting.[7] In the moist tropics, on the other hand, as a result of high transport costs, fuelwood extraction occurs mainly near cities, and industrial roundwood accounts for a much bigger share of total production.[8] In Malaysia, 84 percent of the wood produced goes to timber.[9]

Industrial roundwood production is dominated by a few countries. The United States, Canada, China, and the Russian Federation together produced 50 percent of the total of 1.5 billion cubic meters in 1995.[10] While tropical forests receive the most public attention, temperate and boreal forests actually account for 83 percent of the total volume of industrial roundwood produced.[11]

Currently, 45 percent of all industrial roundwood goes to pulp, paper, and board.[12]

It is estimated that pulp production alone will soon consume more than half the world's annual commercial timber cut.[13] Industrial countries continue to dominate paper production despite the recent proliferation of pulp mills in many developing countries. For example, Africa's total paper production increased by 70 percent in the eighties, yet it remains only 3.5 percent the level in the United States.[14] Industrial nations as a whole accounted for 83 percent of paper production in 1991.[15]

Paper consumption continues to rise everywhere. The world consumes five times as much paper today as it did in 1950 and will likely use twice as much again by 2010.[16] Usage in industrial countries is more than 10 times that in developing countries.[17] The growing market for pulp and paper is the main driving force behind development of monoculture timber plantations that often replace natural forests. The print run of the Sunday *New York Times* alone requires 75,000 trees.[18]

In the future, well-managed forests and plantations could supply a sustainable source of wood without causing unacceptable environmental damage. In an effort to promote this through labeling wood products, the Forest Stewardship Council—an international body endorsed by environmental groups—has certified 5.5 million cubic meters of forest products as sustainably produced from 3.1 million hectares of forest worldwide.[19]

The role of the international timber trade may appear minimal because exports account for only 7 percent of all industrial roundwood in the marketplace.[20] But these global figures mask important regional differences. Nearly 90 percent of global trade is from the temperate and boreal forests in the North, where, according to the World Wildlife Fund, timber trade is now the primary cause of natural forest loss.[21] Domestic consumption of timber in tropical countries is likely to increase due to growth of population and incomes. The level and relative contribution of the tropical timber trade will probably grow as well, as many developing countries become net timber importers.

WORLD PRODUCTION OF ROUNDWOOD, 1950–95

YEAR	TOTAL (mill. cubic meters)
1950	1,421
1955	1,496
1960	1,753
1965	2,214
1966	2,240
1967	2,266
1968	2,303
1969	2,340
1970	2,390
1971	2,431
1972	2,436
1973	2,511
1974	2,431
1975	2,579
1976	2,687
1977	2,709
1978	2,793
1979	2,877
1980	2,927
1981	2,931
1982	2,926
1983	3,038
1984	3,146
1985	3,180
1986	3,268
1987	3,343
1988	3,401
1989	3,459
1990	3,506
1991	3,397
1992	3,405
1993	3,416
1994	3,440
1995	3,461

SOURCES: FAO, *Forest Products Yearbook* (various years); Mafa Chipeta, FAO, discussion with author, 30 December 1996.

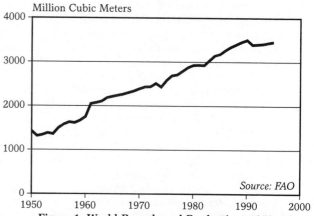

Figure 1: World Roundwood Production, 1950–95

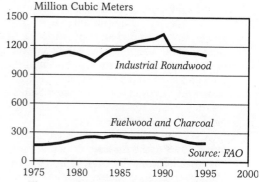

Figure 2: Industrial Countries: Production of Fuelwood and Charcoal versus Industrial Roundwood, 1975–95

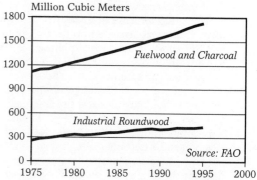

Figure 3: Developing Countries: Production of Fuelwood and Charcoal versus Industrial Roundwood, 1975–95

Storm Damages Set Record   Christopher Flavin

Economic damages from weather-related disasters reached a record $60 billion in 1996.[1] (See Figure 1.) This continues a trend of soaring weather-related losses in the 1990s. So far in the decade, these losses have exceeded $200 billion—four times the total losses for the entire 1980s.[2]

Floods, hurricanes, and wildfires were among the climate-related catastrophes that contributed to the record losses in 1996. Severe flooding in China represented the greatest single loss, totalling $26 billion.[3] These floods, which hit both the Yellow and the Yangtze river basins in July, killed 2,700 people and drove 2 million from their homes.[4] Damage to crops in some areas was severe.

A series of cyclones in the northwestern Pacific Ocean contributed to the damages in China as well as in Taiwan, the Philippines, and Viet Nam.[5] Other notable weather disasters of 1996 included an active Atlantic hurricane season, which caused extensive damage in Cuba, the Dominican Republic, and Puerto Rico and culminated in Hurricane Fran, which hit the southeastern United States and resulted in insured losses of $1.6 billion—most of it from flooding.[6]

A severe cyclone in India killed 2,000 people in 1996, and a massive drought-related bush fire in Mongolia covered more than 100,000 square kilometers.[7] In late December 1996 and early January 1997, a series of unusually strong Pacific storms caused at least $1.6 billion of flood damage in California alone.[8]

The sharp increase in losses from weather-related disasters appears to reflect a combination of factors. One is the rapid increase in housing and industry in high-risk areas such as coastlines and floodplains. The second, less certain factor is human-induced climate change, which may increase the frequency and severity of weather disasters. Thomas Karl, a top climatologist with the National Oceanic and Atmospheric Administration in the United States, reports that in recent decades the number of blizzards and heavy rainstorms has jumped 20 percent in the United States, throwing conventional storm planning techniques into disarray.[9] "Hundred-year events are becoming more frequent now," says Karl.[10]

According to a 1997 statement of the Munich Reinsurance Company, one of the world's largest, "Changes in the environment and climate are leading to a greater probability of new extremes in temperatures, precipitation, water levels, and wind velocities.... That is why the Munich Re has long been pleading for speedy and comprehensive measures to be taken with a view of man-made changes in the environment."[11]

Insurance industry losses from weather-related disasters in 1996 came to $9 billion, the third-highest ever recorded—though well short of the $22 billion in 1992, the year of Hurricane Andrew.[12] (See Figure 2.) Insurance companies were spared by the fact that most of the damages in 1996 came in developing countries, where most people and businesses cannot afford insurance. In China, just $400 million—1.5 percent—of the total losses were insured.[13] In addition, many insurance companies have reduced their exposure in areas that are vulnerable to hurricanes and floods in response to their losses in the early 1990s.

In August 1996, 13 large companies formed a new Risk Prediction Initiative to be run by the Bermuda Biological Station for Research.[14] This will allow insurers to work with scientists to develop improved tools for anticipating the scale of future disasters.

In July 1996, for the first time, a large delegation of insurers attended the Conference of the Parties to the Convention on Climate Change in Geneva. Under the auspices of the U.N. Environment Programme, some 60 insurers—including multibillion-dollar companies such as General Accident, Swiss Reinsurance Company, and Sumitomo Marine & Fire Company—signed a statement calling on governments to reduce substantially the emissions of climate-altering greenhouse gases.[15]

ECONOMIC LOSSES FROM WEATHER-
RELATED NATURAL DISASTERS
WORLDWIDE, 1980–96

YEAR	OVERALL LOSSES (bill. dollars)
1980	1.5
1981	7.8
1982	2.1
1983	6.2
1984	2.3
1985	5.0
1986	6.7
1987	9.6
1988	3.2
1989	9.7
1990	15.0
1991	27.0
1992	36.0
1993	22.5
1994	22.5
1995	38.5
1996 (prel)	60.0

YEAR	INSURED LOSSES (bill. dollars)
1980	0.1
1981	0.4
1982	1.0
1983	2.9
1984	1.0
1985	2.0
1986	0.2
1987	4.3
1988	0.8
1989	4.5
1990	10.0
1991	8.0
1992	22.5
1993	5.5
1994	1.8
1995	9.0
1996 (prel)	9.0

SOURCE: Gerhard A. Berz, Münchener
Rückversicherungs-Gesellschaft, press
release (Munich, Germany: 23 December
1996).

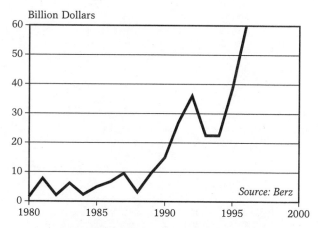

Figure 1: Economic Losses from Weather-Related
Natural Disasters Worldwide, 1980–96

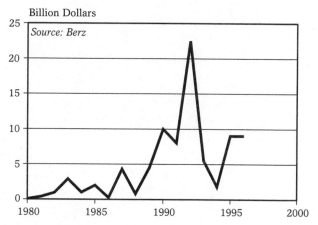

Figure 2: Insured Losses from Weather-Related
Natural Disasters Worldwide, 1980–96

Transportation
Trends

Automobile Fleet Expands   Seth Dunn

Global production of automobiles edged up to 36.1 million in 1996, near the 1990 peak of 36.3 million (see Figure 1); this swelled the world car fleet 2 percent, to an estimated 496 million.[1] (See Figure 2.) This 10-million increase in the number of cars is well below the annual rise of 21 million in 1990.[2] Still, while human population doubled since 1950, the number of cars increased nearly tenfold.[3] (See Figure 3.)

World car sales, meanwhile, set a new record of 36.2 million. In the United States, Japan, and Western Europe—where the fleet grew 2 percent, to 333 million cars—demand and output are flattening or dropping.[4] But they are rising steeply in many parts of Eastern Europe, Latin America, and Asia.

Eastern Europe's car fleet grew 4 percent, to 37 million, as sales jumped 10 percent; Russia's car numbers expanded by 6 percent, to 14 million.[5] Sales in the Czech Republic and Poland leapt 25 and 29 percent.[6]

Latin America's fleet size also rose 4 percent, adding 1.4 million cars in 1996, for a total of 33.4 million.[7] Annual sales, rising 3 percent, have now reached 2 million cars.[8] In Brazil, which alone added 1 million cars in 1996, both production and sales have nearly doubled since 1992.[9]

The most dramatic growth in car numbers occurred in Asia outside Japan, where the fleet size rose 15 percent, to 19.5 million.[10] Output increased 13 percent to 3.2 million; sales, however, rose only 2.8 percent, leading analysts to lower their sales projections.[11] Korea's car fleet jumped 25 percent, to 6.4 million, with sales up 8 percent.[12] In China, output rose 2 percent and has grown nearly tenfold since 1991.[13] Demand, though, was lower than in 1993, forcing plants to scale back production and to stockpile more than 200,000 unsold cars.[14]

Asian governments and international automakers continue to promote car production. Nine of the world's automakers have entered India's market, boosting production 16 percent and sales 17 percent.[15] Korean carmakers are now producing for European as well as Asian markets, and Malaysia and Indonesia plan to launch domestic car indus-

tries.[16] Several major carmakers have targeted Thailand as an Asian production base.[17]

Automakers are also especially interested in China; the government there, intent on having carmaking as a "pillar industry" and pushing the country's car fleet from 2.7 million to 22 million by 2010, has lowered tariffs to encourage domestic production.[18]

Analysts project a doubling of the world fleet in 25 years, yet the social and environmental costs of car use are escalating.[19] Congestion in the United States accounts for $100 billion in wasted fuel, lost productivity, and rising health costs.[20] In Bangkok, cars spend the equivalent of 44 days a year stopped in traffic, costing $2.3 billion in lost time.[21] Air pollution in Mexico City, among the world's worst, has health impacts estimated at $1.5 billion per year.[22] Globally, cars are responsible for one third of oil consumption and more than 15 percent of greenhouse gas emissions.[23]

Efforts to restrict car use have multiplied. Mexico City, Athens, and Singapore control usage through a system based on license numbers.[24] Highway plans in Tokyo, road-building projects in the United Kingdom, and urban sprawl in Portland, Oregon, have all been halted to avoid increasing dependence on cars.[25] A number of cities in France and Italy are prohibiting cars from urban centers; Thailand's foreign minister has proposed banning new cars entirely in Bangkok between 1997 and 2001.[26]

Alternatives to dependence on automobiles are found in industrial and developing countries. In Copenhagen, one third of commuters now bicycle to work; in Curitiba, Brazil, 70 percent of commuters travel by bus.[27] In Asia, nonmotorized vehicles are the most popular form of transportation.[28]

Transport planning nevertheless continues to focus on motorization. China funds extensive road-building projects but has recently cancelled mass transit projects; in Bangkok, mass transit receives no public funding.[29] Highway-related lending accounts for 60 percent of World Bank transportation lending.[30] Such policies perpetuate the unrestrained growth of the world automobile fleet.

WORLD AUTOMOBILE PRODUCTION AND
FLEET, 1950–96

YEAR	PRODUCTION (million)	FLEET (million)
1950	8	53
1955	11	73
1960	13	98
1965	19	140
1966	19	148
1967	19	158
1968	22	170
1969	23	181
1970	23	194
1971	26	207
1972	28	220
1973	30	236
1974	26	249
1975	25	260
1976	29	269
1977	31	285
1978	31	297
1979	31	308
1980	29	320
1981	28	331
1982	27	340
1983	30	352
1984	31	365
1985	32	374
1986	33	386
1987	33	394
1988	34	413
1989	36	424
1990	36	445
1991	35	456
1992	35	470
1993	34	470
1994	35	480
1995	36	486
1996 (prel)	36	496

SOURCES: American Automobile Manufacturers
Association; DRI/McGraw-Hill.

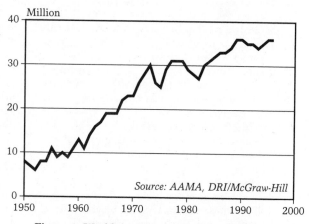

Source: AAMA, DRI/McGraw-Hill

Figure 1: World Automobile Production, 1950–96

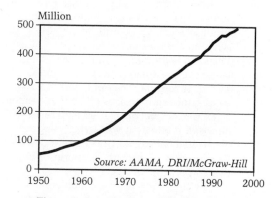

Source: AAMA, DRI/McGraw-Hill

Figure 2: World Automobile Fleet, 1950–96

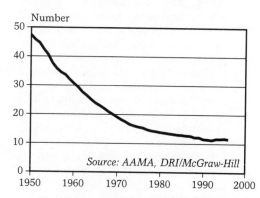

Source: AAMA, DRI/McGraw-Hill

Figure 3: People Per Automobile, 1950–96

Bicycle Production Still Rising

Bicycle production was up in 1995, the last year for which extensive data are available.[1] Some 109 million bicycles were produced that year, 1.9 percent more than in 1994.[2] (See Figure 1.) Steady growth in output this decade, however, has also swollen inventories, leading manufacturers to worry whether production increases can be sustained in the next few years.[3]

Sixty-five nations worldwide produce bikes.[4] China dominates, with output of 41 million units in 1995—some 40 percent of reported world production.[5] The next largest producer, the European Union (EU), built only a third as many bikes as China, some 13 million units.[6] India, a rising bicycle power, is close behind, having manufactured 11.5 million bikes in 1995.[7] (See Figure 2.)

Asian dominance of global production has increased in the 1990s, in spite of a slump in output in Japan and South Korea.[8] Reporting Asian nations accounted for 69 percent of global production in 1990; by 1995, their share rose to 73 percent.[9] China, India, Taiwan, Indonesia, Thailand, and Malaysia saw production increases of from 29 percent to 185 percent between 1990 and 1995, while European production was largely stagnant.[10] The United States and Brazil also did well this decade, with growth in output of more than 50 percent.[11]

Growth in bicycle production is influenced by many factors, including changes in technology. The recent emergence of electric bicycles in wealthy nations, for example, could expand the vehicle's commuting potential by lengthening its range and reducing pedaling effort.[12] One manufacturer expects 1997 sales of electric models in Japan alone to reach 100,000 units, double the level of 1996.[13] Possibly of greater importance is a simple $15 bicycle developed in 1996 by a California firm.[14] The inexpensive design—made from recycled plastics and sheet metal, with the pedals attached to the front wheel, eliminating the chain—could make cycling more accessible in developing countries, where bicycle purchases can require several months' wages.[15]

Perhaps the greatest influence on cycling levels is government policy. Bicycle rickshaws in Jakarta, for example, numbered more than 100,000 in the early 1970s, but their ranks were reduced to some 30,000 in 1988 after image-conscious officials banned them.[16] Even where the official attitude is indifference rather than hostility, cycling can suffer. In U.S. cities, unsafe biking conditions—the result of slighting bike needs in transportation planning—is one of the three main reasons people give for not cycling.[17] The danger to cyclists is reflected in employer-paid insurance rates for New York's bike messengers, which rose from 1 percent of payroll a few years back to 6 percent in 1996.[18]

In contrast, cycling prospers when nurtured. Copenhagen, with some 300 kilometers of bikeways, provides 1,000 bikes for free citizen use around the city.[19] Businesses sponsor the bikes—in exchange for advertising space on them—while the city maintains the fleet.[20] As a result, cycling in Copenhagen accounts for 20 percent of daily trips, compared with less than 1 percent in the United States.[21] Other cities in Europe and Japan boost ridership through "bike and ride" programs, which conveniently link cycling with public transportation.[22]

Inspired by these successes, Peru's capital is constructing 51 kilometers of bikeways and redesigning 35 kilometers of roads to accommodate bikes, in an effort to raise bike use from 2 percent of trips to more than 10 percent.[23] Lima also plans to provide 17,000 loans for bicycle purchases for the poor over a two-year period.[24] In the EU, where transport now uses more energy than industry does, officials in 1996 included bicycles for the first time as an integral part of the EU transportation plan.[25] And the United Kingdom announced in 1996 that bike-friendly policies would be used to double bicycle use by 2002, and to double it again by 2012.[26] Widespread adoption of similar policies could raise the profile of the bicycle in cities around the world.

WORLD BICYCLE PRODUCTION,
1950–95

YEAR	PRODUCTION (million)
1950	11
1955	15
1960	20
1965	21
1966	22
1967	23
1968	24
1969	25
1970	36
1971	39
1972	46
1973	52
1974	52
1975	43
1976	47
1977	49
1978	51
1979	54
1980	62
1981	65
1982	69
1983	74
1984	76
1985	79
1986	84
1987	98
1988	105
1989	95
1990	92
1991	100
1992	102
1993	106
1994	107
1995	109

SOURCES: United Nations, *The Growth of World Industry 1969 Edition*, Vol. II, *Yearbooks of Industrial Statistics 1979 and 1989 Editions*, Vol. II, and *Industrial Commodity Statistics Yearbook 1994*; *Interbike Directory 1997*.

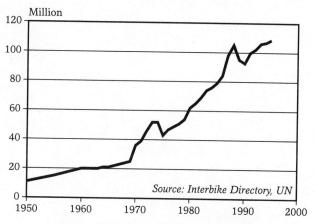

Figure 1: World Bicycle Production, 1950–95

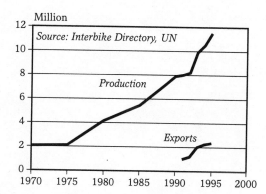

Figure 2: Bicycle Production and Exports
In India, 1970–95

Social
Trends

Population Increase Slows Slightly   Jennifer D. Mitchell

World population increased by 80 million people in 1996, bringing the global total to 5.77 billion.[1] (See Figures 1 and 2.) Most of this growth—about 98 percent—occurred in developing countries.[2]

While the overall number of people continues to increase, the annual rate of growth has been slowly dropping from its record high of 2.2 percent in 1963; in 1996, it fell to 1.4 percent.[3] (See Figure 3.) The significance of this decline is limited, however, due to the expanding base population: the 2.2-percent growth rate in 1963 yielded 69 million more people, but the 1.4-percent growth rate in 1996 produced an additional 80 million.[4]

Although world population continues to increase, it is growing more slowly than expected.[5] Two new factors—updated census data indicating lower fertility rates in many Asian countries, and the incorporation of AIDS mortality data for seven additional countries—resulted in revisions to earlier population data.[6] Demographers originally estimated that 86–90 million people had been added to world population each year since 1986.[7] The new data, however, reduced these figures by 1–7 million a year over the past decade, so the world total in 1996 was 41 million fewer people than expected.[8]

With these revisions included, the annual addition dropped from a peak of 87 million in 1990 to 80 million in 1996.[9] Ideally this decrease would be due to couples willingly having fewer children, but unfortunately part of the decline was due to increased mortality in several regions.

Between 1990 and 1995, average life expectancy in Russia dropped sharply, from 64 to 57 for men and from 74 to 70 for women, due largely to cardiovascular disease, accidents, murder, suicide, and excessive consumption of alcohol.[10] Life expectancy has also declined in the other 14 former Soviet republics since 1990.[11] Similarly, in sub-Saharan Africa the rate of growth has partially been kept in check by rapidly increasing mortality due to AIDS; fertility rates in this region have declined only slightly.[12]

Progress in lowering fertility rates is never-theless being made. Worldwide, the average number of children born to a woman in her lifetime (the fertility rate) dropped by more than one child per woman between 1985 and 1996.[13] During this period, the total fertility rate fell 26 percent in India, 30 percent in Brazil, and 35 percent in Bangladesh.[14]

Despite these declines, the U.S. Census Bureau estimates that the annual number of births worldwide will remain over 132 million into the next century due to the large number of women of reproductive age in developing countries.[15] For example, India's total fertility rate dropped from 4.3 in 1985 to 3.2 in 1996, yet due to its already massive population size, close to 25 million babies were born there last year.[16] Even with this drop in fertility, India's population is expected to reach 1 billion by 2000.[17]

Falling fertility rates are partially due to better access to contraceptives and reproductive services. In developing countries today, five times as many couples are using contraceptives as in the 1960s, but contraceptives are still unavailable to more than 120 million women worldwide.[18] Access to health care, family planning, and education will need to be increased just to keep up with the growing number of females of reproductive age.[19]

Although some countries, such as the Netherlands, increased their support for family planning programs after the 1994 International Conference on Population and Development in Cairo, U.S. support plummeted from $582 million in 1995 to $76 million in 1996.[20] Analysts caution that "donor fatigue" could reverse recent progress in lowering fertility rates.[21]

Rapid population growth undermines developing countries' capacity to address urgent social, economic, and environmental problems. In the words of Dr. Rafiq Zakaria, delegate to the United Nations Commission on Population and Development, for many countries population growth is "a question of survival because it eats into economic progress to an extent that, unless it is rapidly checked, we get pushed two steps backward every time we advance one step forward."[22]

WORLD POPULATION, 1950–96

YEAR	TOTAL (billion)	ANNUAL ADDITION (million)
1950	2.556	37
1955	2.780	51
1960	3.039	42
1965	3.345	69
1966	3.415	70
1967	3.485	70
1968	3.556	72
1969	3.631	75
1970	3.706	75
1971	3.783	77
1972	3.861	77
1973	3.937	77
1974	4.013	76
1975	4.087	74
1976	4.160	72
1977	4.233	73
1978	4.305	73
1979	4.381	76
1980	4.458	77
1981	4.534	77
1982	4.615	80
1983	4.695	80
1984	4.774	79
1985	4.855	81
1986	4.937	82
1987	5.023	86
1988	5.109	87
1989	5.194	85
1990	5.282	87
1991	5.366	84
1992	5.448	82
1993	5.529	81
1994	5.610	80
1995	5.691	81
1996 (prel)	5.772	80

SOURCE: U.S. Bureau of the Census, *International Data Base*, electronic database, Suitland, MD, 15 May 1996.

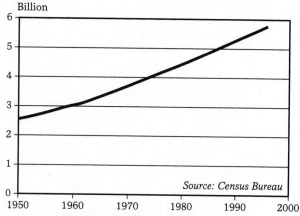

Figure 1: World Population, 1950–96

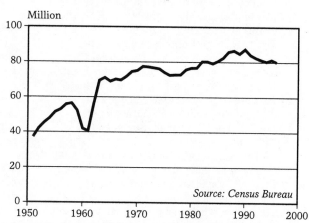

Figure 2: Annual Addition to World Population, 1950–96

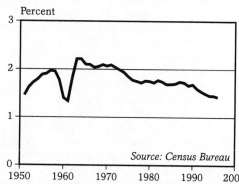

Figure 3: Average Annual Growth Rate of World Population, 1950–96

Refugee Population Has Rare Decline   Hal Kane

The number of people in the world deemed by the United Nations to be in need of and eligible for assistance as refugees dropped by 1.3 million individuals in 1996.[1] (See Figure 1.) It was the largest reduction since the end of World War II—and one of the few times since then that any drop has occurred at all.[2]

The world's refugee population grew steeply since the late 1970s, and particularly steeply since 1992, to reach an all-time high of 27.4 million people at the beginning of 1995.[3] This group includes people classified by the United Nations as refugees, recent returnees, "others of concern," and some internally displaced.

Since 1951, refugees have been defined as people who have to remain outside their country because of a well-founded fear of persecution or other danger.[4] The people who fit this strict definition number 13.2 million.[5] The U.N. High Commission for Refugees is also helping 3.3 million "returnees"—former refugees, most of whom have gone home within the past two years.[6] Most returnees receiving assistance are in Afghanistan, Cambodia, Ethiopia, Iraq, Mozambique, Myanmar, Somalia, and Tajikistan.[7]

The category "others of concern"—4.9 million people—includes war victims in the former Yugoslavia who were not displaced but still receive help, and people in the Russian Federation who are moving back to regions that Soviet authorities had forced them out of earlier in history.[8] It also includes rejected Vietnamese and Laotian asylum-seekers in Southeast Asia who need help, and a few other groups of people.[9]

By many estimates, at least 20 million people, and likely more than 30 million, are "internally displaced"—living in a refugee-like situation within their own countries.[10] The United Nations can only assist them if their governments allow the help. Last year, it aided 4.7 million such people, but that means that most internally displaced people did not receive international assistance.[11]

The main drop in people receiving assistance occurred in Africa, which registered 2.7 million fewer such people at the beginning of 1996.[12] (See Figure 2.) One million of these were refugees no longer needing aid because of repatriations.[13] In addition, about 1 million returnees from past years are no longer receiving U.N. assistance, and the number of internally displaced recipients fell by a half-million.[14]

The African repatriations may be temporary, however, since Rwanda is not stable, and since new emergencies are beginning to arise in other places as well.[15] During 1996, Burundi continued to show signs of war, and Zaire is now disintegrating, which could create a new group of refugees.[16] Even after recent repatriations, Rwandans still constitute the second largest group of refugees in the world—1.7 million people.[17]

Asia had a slight net reduction in its refugee population. Some 348,000 Afghans went home from India, Iran, and Pakistan between January 1, 1995, and January 1, 1996, while some 66,000 Burmese returned to Myanmar.[18] Both these countries continue to have problems, however, and there is no guarantee of lasting peace. Currently, Afghanistan leads the world as a source of refugees, at 2.35 million.[19]

Europe produced few refugees between World War II and the early 1990s, but the continent now has as many people receiving refugee assistance as Asia does, and almost as many as Africa.[20] Some 1.33 million refugees remained outside of Bosnia, and more than a million remained displaced inside that new country's borders as of January 1, 1996, an all-time high.[21]

All these categories are for people affected by human conflict. But research suggests that the number of people affected by non-conflict disasters, such as floods and earthquakes, has been increasing.[22] So far during the 1990s, the number of people needing such humanitarian assistance has ranged from 100 million (in 1992) to 350 million (in 1991).[23] These people are not refugees, though they may still need help and they may compete with refugees for assistance.

REFUGEES RECEIVING U.N.
ASSISTANCE, 1961–96

YEAR	TOTAL (million)
1961	1.4
1962	1.3
1963	1.3
1964	1.3
1965	1.5
1966	1.6
1967	1.8
1968	2.0
1969	2.2
1970	2.3
1971	2.5
1972	2.5
1973	2.4
1974	2.4
1975	2.4
1976	2.6
1977	2.8
1978	3.3
1979	4.6
1980	5.7
1981	8.2
1982	9.8
1983	10.4
1984	10.9
1985	10.5
1986	11.6
1987	12.4
1988	13.3
1989	14.8
1990	14.9
1991	17.2
1992	17.0
1993	19.0
1994	23.0
1995	27.4
1996 (prel)	26.1

SOURCE: United Nations High
Commissioner for Refugees, various
data series.

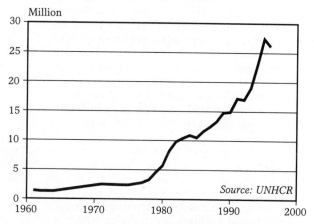

Figure 1: Refugees Receiving U.N. Assistance, 1961–96

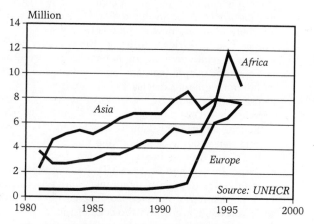

Figure 2: Refugees Receiving U.N. Assistance in Asia,
Africa, and Europe, 1981–96

HIV/AIDS Pandemic Broadens Reach · Jennifer D. Mitchell

Between the late 1970s—when the human immunodeficiency virus pandemic began—and the end of 1996, an estimated 36.2 million people worldwide were infected with HIV; 12.5 million of these individuals have acquired full-blown AIDS—the fatal syndrome believed to develop about a decade after HIV infection.[1] (See Figures 1 and 2.)

Despite proven methods of prevention, the number of people infected each year continues to increase. In 1996, a record 5.6 million people contracted HIV.[2] The number of AIDS deaths also reached a record high in 1996: 1.7 million.[3] As serious as the current situation is, the future may be even bleaker as the number of AIDS deaths continues to grow. Of the 36.2 million infected with HIV, one third have died of AIDS thus far.[4]

Sub-Saharan Africa—with just 10 percent of the world's population—accounts for 63 percent of the current HIV infections.[5] The impact of HIV/AIDS in this region is severe. Yet there are more new infections each year in Asia than in Africa.[6] Although the epidemic there is relatively new, India has more HIV-positive people than any other country, and demographers expect it to become the center of the disease by the end of this decade.[7]

Recent surges in previously unaffected populations provide clear warning that HIV/AIDS is broadening its reach. In Ukraine, Russia, and Belarus, HIV has rapidly expanded due to poverty, ignorance, and a large population of intravenous drug users.[8] In Nikolayev, Ukraine, on the edge of the Black Sea, the share of HIV-infected people among drug users exploded from 1.7 percent in 1995 to 56.5 percent just 11 months later.[9] New outbreaks are also occurring in other parts of Eastern Europe, Viet Nam, Cambodia, and China.[10]

In addition to expanding its geographical range, HIV/AIDS is extending its reach within society. The number of women and children infected with HIV/AIDS is growing rapidly. Approximately 3,500 women are infected each day, and 25–35 percent of all infants born to HIV-infected women become infected before or during birth, or through breast-feeding.[11]

And the impacts of AIDS extend beyond those who have the virus. In Kenya, Rwanda, Zambia, and Uganda, close to 1 million children have been orphaned because their parents died of AIDS.[12]

As the pandemic covers more ground, it is fragmenting and becoming many epidemics with diverging modes of transmission, affected populations, and rates of spread. In Southeast Asia and sub-Saharan Africa, it is spread mainly by heterosexual intercourse; in North America and Western Europe, by homosexual intercourse; and in Eastern Europe, by drug use.[13]

But over time these patterns are also changing. At the start of the epidemic, most infections in Latin America, for example, were due to homosexual or bisexual contact; currently, however, 75 percent of new infections are the result of heterosexual contact.[14]

The new Joint United Nations Programme on HIV/AIDS and a group of 60–70 specialists from the affected countries met to review these regional patterns at the 11th International Conference on AIDS in Vancouver, in July 1996.[15] Meetings in the Philippines, Peru, and Côte d'Ivoire in 1997 will explore these patterns further.[16] Recognizing the different paths of transmission helps target preventative measures toward "at-risk" populations.

Recent work has shown that HIV infection rates can be cut drastically at little cost by treating and preventing other sexually transmitted diseases (STDs). An intensive program to diagnose and treat STDs in Mwanza, Tanzania, reduced the rates of HIV infection by more than 40 percent.[17] Similarly, needle exchange programs may be effective in regions where the main mode of transmission is intravenous drug use.[18]

For those already infected, new drugs (protease inhibitors) greatly reduce levels of HIV when combined with AZT and other antiviral drugs; but at $20,000 per year, most patients cannot afford these treatments.[19] For most people, therefore, the best hope lies in education and prevention.

ESTIMATES OF CUMULATIVE
HIV/AIDS CASES WORLDWIDE,
1980–96

YEAR	HIV INFECTIONS (million)
1980	0.2
1981	0.6
1982	1.1
1983	1.8
1984	2.7
1985	3.9
1986	5.3
1987	6.9
1988	8.7
1989	10.7
1990	13.0
1991	15.5
1992	18.5
1993	21.9
1994	25.9
1995	30.6
1996 (prel)	36.2

YEAR	AIDS CASES (million)
1980	0.0
1981	0.1
1982	0.1
1983	0.1
1984	0.2
1985	0.4
1986	0.7
1987	1.1
1988	1.6
1989	2.3
1990	3.2
1991	4.2
1992	5.5
1993	6.9
1994	8.5
1995	10.4
1996 (prel)	12.5

SOURCE: Global AIDS Policy Coalition,
Harvard School of Public Health,
Cambridge, MA, discussion with author,
24 January 1997.

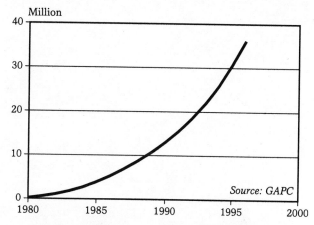

Figure 1: Estimates of Cumulative HIV Infections Worldwide, 1980–96

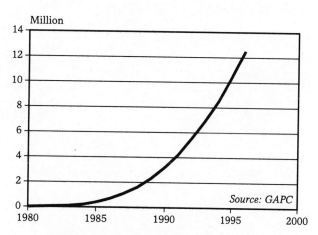

Figure 2: Estimates of Cumulative AIDS Cases Worldwide, 1980–96

Military

Trends

U.N. Peacekeeping Declines Sharply
Michael Renner

Expenditures for United Nations peacekeeping operations dropped to about $1.8 billion in 1996, down 44 percent from the previous year.[1] (See Figure 1.) This was the lowest amount since 1993, equivalent to 20 hours' worth of world military spending.[2] The number of military troops, observers, and civilian police involved in peacekeeping declined from about 53,000 in November 1995 to 25,649 in November 1996.[3] (See Figure 2.)

The sharp drop in expenditures and personnel is the result primarily of the termination of two large missions in the former Yugoslavia, the end of the U.N. presence in Rwanda, and the downsizing of operations in Haiti.[4] Also, 1996 was the first year since 1990 during which no entirely new operations were initiated.[5]

The single largest of 16 remaining U.N. operations is the Angola Verification Mission (UNAVEM III), established to help implement the 1991 and 1994 peace accords in that country. UNAVEM III deploys about 7,200 military personnel and civilian police and costs $323 million per year.[6]

With the U.S.-brokered Dayton Peace Agreement for Bosnia, U.N. peacekeeping in the former Yugoslavia was sidelined by a NATO-led force, IFOR.[7] During 1996, IFOR spent $5 billion deploying some 60,000 troops; a scaled-down force of 31,000 troops has been in place since December 1996, which is still larger than all current U.N. operations combined.[8]

A scaled-back U.N. presence also remains in Bosnia, along with two operations in pockets of Croatia and a small force in Macedonia.[9] Together, these U.N. missions cost about $400 million in 1996, a far cry from the almost $1.8 billion spent by the United Nations in the former Yugoslavia in 1995.[10]

From 1948 to 1996, a total of 42 U.N. missions were initiated to help settle 32 distinct conflicts, at a cumulative cost of $17.2 billion.[11] Since 1948, more than 750,000 individuals from 110 nations have served as peacekeepers in military and police functions.[12] (In addition, tens of thousands of people have served in civilian posts since the early 1990s.) More than 1,400 peacekeepers have died while on duty.[13]

Peacekeeping is handicapped by the U.N.'s severe financial crisis. At $1.6 billion (as of December 31, 1996), outstanding contributions to the peacekeeping budget now almost equal annual expenditures.[14] (See Figure 3.) Although all but 16 of the 185 member states owe some dues, the top 5 debtors together account for 90 percent of all arrears.[15] The United States alone is responsible for $926 million, or 57 percent.[16]

Congressional hostility makes it unclear whether the United States will pay its dues and clear its past arrears, as the Clinton administration has repeatedly proposed.[17] Kofi Annan, the new Secretary-General, is well regarded, but it remains to be seen whether he can persuade Congress of the value of a strong United Nations.[18]

Several efforts are under way to improve U.N. capacity to carry out effective peacekeeping operations. A "Lessons-Learned Unit" was established in April 1995 to explore more systematically how past experience can help improve future efforts.[19] Since 1993, the United Nations has been working to establish standby arrangements, under which member governments would designate specific military units and equipment that they might make available for future peacekeeping service. By January 1997, 62 countries had expressed willingness to participate, but only Austria, Denmark, Ghana, Jordan, and Malaysia had formalized their offers.[20]

After rapid growth and an equally quick retrenchment, peacekeeping has entered a period of consolidation. It is unclear how long the Bosnia and Croatia missions will extend; the Haiti operation is to close by July 1997.[21] A human rights mission to Guatemala was transformed in early 1997 into a military observer mission to oversee implementation of a peace accord; a similar operation may soon be set up in Sierra Leone.[22] And a small non-U.N. force is to be sent into the Central African Republic to supervise a cease-fire there.[23] But no large or dramatic new operations are likely to be launched any time soon.

U.N. PEACEKEEPING OPERATION EXPENDITURES, 1986–96

YEAR	EXPENDITURE (mill. dollars)
1986	242
1987	240
1988	266
1989	635
1990	464
1991	490
1992	1,767
1993	3,055
1994	3,357
1995	3,281
1996 (prel)	1,840

SOURCES: U.N. Department of Peacekeeping Operations, discussion with author, 20 December 1995; "Peace-Keeping Operations, Expenditures (All Missions)," < http://www.un.org/Depts /DPKO/yir96/allexp.jpg >, viewed 18 March 1997.

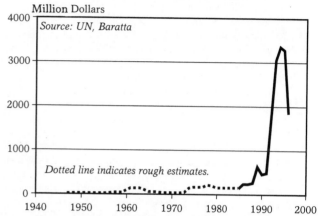

Million Dollars

Source: UN, Baratta

Dotted line indicates rough estimates.

Figure 1: U.N. Peacekeeping Expenditures, 1947–96

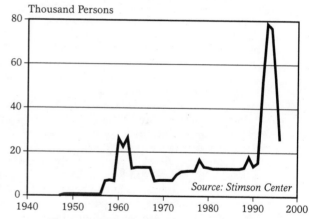

Thousand Persons

Source: Stimson Center

Figure 2: U.N. Peacekeeping Troops, 1947–96

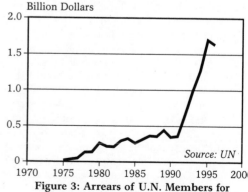

Billion Dollars

Source: UN

Figure 3: Arrears of U.N. Members for Peacekeeping Expenses, 1975–96

The ranks of the world's armed forces declined to 23 million in 1995, the most recent year for which data are available.[1] This is a 6-percent fall from the 24.6 million in 1994, and a 20-percent cut from the peak of 28.7 million in 1988.[2] (See Figure 1.) The decline is due to the end of the cold war, the termination of several long-lasting conflicts in the Third World, and budgetary difficulties in many countries.

The U.S. Arms Control and Disarmament Agency (ACDA), the source of these data, combines active-duty military personnel in its estimates with paramilitary forces that "resemble regular units in their organization, equipment, training, or mission."[3]

Using a broader definition, the International Institute for Strategic Studies (IISS) in London puts paramilitary forces at 7.1 million (in addition to estimating 22.6 million regular armed forces).[4] And IISS estimates the number of reservists, a category not covered by ACDA, at 39.3 million.[5] There are roughly 4 soldiers for every 1,000 civilians in the world, but counting paramilitary forces and reservists, the ratio grows to 13 per 1,000.[6]

Training, equipment, and readiness of troops vary enormously. Some of the richest countries spend as much as $100,000 to $300,000 annually per soldier, while the poorest spend $1,000 or less.[7] In Russia, due to appalling conditions, more than half of all conscripts now evade the military draft.[8]

About 500,000 soldiers were stationed outside their own countries in mid-1996, including roughly 100,000 soldiers deployed in a variety of U.N. and non-U.N. peacekeeping operations.[9] The United States has 238,000 troops abroad.[10] Russia maintains some 34,000 troops on foreign soils, mostly in the former Soviet republics—a fraction of the roughly 700,000 soldiers that the Soviet Union used to have outside its borders.[11]

Tallies of armed forces usually do not include armed opposition groups—largely because reliable information on them is hard to come by. Data compiled by IISS for 27 countries suggest that such forces number at least 410,000.[12] But adding other countries

with ongoing civil wars would raise the total by several tens of thousands.[13]

On a regional basis, East Asia has by far the largest armies (33 percent of the global total), followed by Europe (27 percent), the Middle East (10 percent), South Asia (9 percent), and North America (8 percent). Africa (6 percent) and Latin America (5 percent) have far lower levels of armed forces.[14]

The five largest national armies are found in China, the United States, Russia, India, and North Korea.[15] (See Figures 2 and 3.) Countries without armies are still a rare exception. Iceland is one of them, but as a member of NATO it hosts some 2,000 U.S. troops.[16] Costa Rica abolished its armed forces in 1949; Panama and Haiti followed suit in 1994 and 1996, respectively, after dictatorships were ousted.[17] Congo has requested European assistance in abolishing its army and generating alternative jobs for its soldiers.[18]

Most countries have no intention of abolishing their armies, but many are trimming them (with some switching from conscripts to professional soldiers). Since the mid-1980s, the largest cuts in absolute terms occurred in China, Ethiopia, Iraq, Russia, Viet Nam, and the United States.[19] The U.S. armed forces are at their lowest level since 1950.[20] Argentina, El Salvador, Eritrea, Mongolia, Mozambique, Namibia, and Nicaragua demobilized half or more of their armies.[21] Angola, Liberia, and South Africa are in the midst of large demobilization efforts.[22] And Guatemala's recent peace agreement stipulates that troop levels be cut by 30 percent.[23]

With smaller armies, some military bases are no longer needed. By the year 2000, more than 8,000 military sites, covering more than 1 million hectares, will have been turned over to civilian use worldwide.[24]

While demobilization can free up scarce resources needed for civilian purposes, it is not a process without costs. Particularly in countries emerging from war, demobilized soldiers need substantial assistance to learn new skills, find jobs, and successfully reintegrate themselves into civilian society. That support is often still inadequate.[25]

ARMED FORCES WORLDWIDE,
1967–95

YEAR	WORLD (thousand persons)
1967	22,734
1968	23,451
1969	24,186
1970	24,493
1971	24,735
1972	25,150
1973	25,478
1974	26,675
1975	26,131
1976	25,740
1977	25,846
1978	26,486
1979	26,716
1980	26,795
1981	27,343
1982	27,256
1983	27,109
1984	28,064
1985	28,075
1986	28,484
1987	28,327
1988	28,702
1989	28,588
1990	27,736
1991	25,964
1992	24,596
1993	24,380
1994	24,598
1995	23,038

SOURCE: ACDA, WMEAT Database.

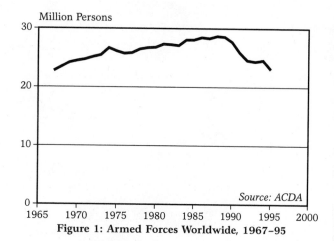

Figure 1: Armed Forces Worldwide, 1967–95

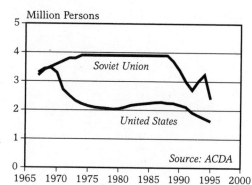

Figure 2: Size of Armies in the United States
and the Soviet Union, 1967–95

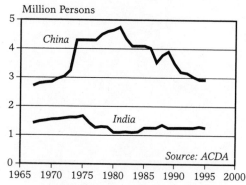

Figure 3: Size of Armies in China
and India, 1967–95

Part **TWO**

Special Features

Environmental Features

Forest Loss Continues   Cheri Sugal

etween 1991 and 1995, the world lost an average of 11.3 million hectares of net forest area annually—an area roughly the size of Honduras.[1] (See Table 1.) Total forest area, not including woodlands, now amounts to some 3.5 billion hectares, 23 percent smaller than in 1700.[2]

Most of the world's deforestation during the first half of this decade was the result of tropical forest loss, which averaged 12.6 million hectares a year.[3] During this time, 4 percent of tropical forests were lost.[4] Despite public attention to the issue of tropical deforestation, the damage has continued unabated from the 1980s: the average annual loss then was 12.8 million hectares.[5]

Within the tropics, Latin America and the Caribbean continue to lose the greatest amount—5.7 million hectares a year in 1991–95, or 3 percent of its forest cover.[6] While Asia and Oceania lost a smaller area than either Africa or Latin America and the Caribbean did, they lost a greater portion of their forests—5 percent.[7] Just a few countries—Bolivia, Brazil, Indonesia, Malaysia, Mexico, Venezuela, and Zaire—together accounted for 50 percent of the tropical forest lost in this five-year period.[8]

The loss of tropical forests may be even greater than these data imply. The figures mask important distinctions in the density of forests. The U.N. Food and Agriculture Organization (FAO) defines deforestation as depletion of crown cover of trees to less than 10 percent.[9] Since most logging in the tropics involves selective cutting, which leaves a ragged residual cover, a logged-over area may not be considered a deforested area. Yet many such areas are deforested in an ecological sense.

The data also mask important distinctions in forest quality. In the tropics, the average annual loss of indigenous tree species, a category labelled "natural forests" by FAO, is greater than the loss of "forests."[10] This is because the latter category includes plantations, which increased by 1.7 million hectares over the first five years of this decade.[11]

Furthermore, because industrial countries do not separate out data on natural forests and plantations, it is impossible to estimate the loss of biological diversity in temperate forests (mainly broad-leaved species in the mid-latitudes) and boreal forests (conifers, aspen, and birch further north). Industrial plantations, most of which contain only one or two tree species, clearly increase net forest area in industrial countries.

The net increase is also due to natural reversion (as marginal land used for agricultural production is abandoned) and active replanting programs. In the former Soviet Union, for example, 4.5 million hectares were reforested in 1980.[12]

Although the area of temperate and boreal forests has remained relatively constant or even increased, the statistics on these areas also mask major degradation in forest quality. In the United States, for instance, although overall forest cover was greater in 1995 than in 1990 (212.5 million versus 209.6 million hectares), 98 percent of all U.S. forest has been logged at least once—with the remaining 2 percent of primary old growth continuing to disappear.[13]

The damage caused by deforestation extends well beyond the forest itself. When plant cover is removed, valuable soils erode, causing siltation to build up in rivers and streams. An estimated 579 million hectares of terrain suffer from soil degradation caused by deforestation.[14] Soils break down organic matter, or waste, into nutrients essential to the growth of green plants, and they soak up precipitation, providing a steady flow of surface water into streams, springs, and aquifers. Deforestation lets rainwater run off the surface—causing floods and droughts downstream.[15]

On a global scale, forests play a vital role in climate regulation by storing carbon. Forests contain between 475 billion and 825 billion tons of carbon.[16] Tropical deforestation releases approximately 1.5 billion tons of carbon to the atmosphere each year—about 19 percent of total carbon emissions worldwide.[17]

Deforestation often results in the loss of important commodities, many of them non-timber forest products. More than half of all prescriptions filled worldwide contain active ingredients originating from wild species—particularly tropical plants.[18] An estimated 80 percent of people living in developing countries rely on traditional medicine for their primary health care needs, and 85 percent of traditional medicines use plant extracts.[19] Yet at current rates of deforestation, 20–75 plant and animal species are lost every day; by 2015, some 6–14 percent of all species are expected to be extinct.[20]

The relative contributions of commercial logging, fuelwood gathering, shifting cultivation, conversion to agriculture, and development projects to the problem of global deforestation are not certain. The role of logging is often underestimated because of its indirectness: logging destroys or fatally injures residual trees and opens up previously intact forests to other causes of deforestation.[21]

In Indonesia, for example, logging accounts directly for approximately 40 percent of deforestation (when defined by the loss of ecosystem functions, not by the percentage of trees).[22] But with another 70,000–100,000 hectares lost due to fires attributable to destructive logging, this practice actually accounts for 50 percent of national deforestation.[23] In Papua New Guinea, logging accounts for one third of the forest loss, but roads connected to logging have opened the forest up to agricultural expansion, which accounts for the bulk of deforestation.[24]

Unfortunately, the vast majority of the world's forests—94 percent—have no official protection from the expanding pressures of human activity.[25] The Convention on Biological Diversity signed at the Earth Summit in 1992 and ratified by 165 nations is developing a position on forests that is consistent with a primary concern for biodiversity conservation and sustainable use of this valuable natural resource.[26]

TABLE 1: GLOBAL FOREST STATUS, BY REGION, 1991–95

REGION	ALL COUNTRIES		DEVELOPING COUNTRIES	
	FOREST AREA, 1995	ANNUAL CHANGE, 1991–95	NATURAL FOREST AREA, 1995	ANNUAL CHANGE, 1991–95
		(million hectares)		
Tropical Regions				
Africa	504.90	− 3.69	502.74	− 3.70
Asia/Oceania	321.67	− 3.21	297.50	− 3.51
Latin America/				
Caribbean	907.39	− 5.69	901.34	− 5.69
Total Tropical	1,733.96	−12.59	1,701.59	−12.90
Nontropical Regions				
Africa	15.34	− 0.05	12.71	− 0.05
Asia/Oceania	243.20	− 0.21	123.16	− 0.66
Latin America/				
Caribbean	42.65	− 0.12	40.93	− 0.12
Europe	145.99	+ 0.39	—	—
Former Soviet Union	816.17	+ 0.56	—	—
North America	457.09	+ 0.76	—	—
Total Nontropical	1,720.42	+ 1.32	176.80	− 0.83
Total Forest Area	3,454.38	−11.27	1,878.39	−13.73

SOURCE: U.N. Food and Agriculture Organization, *State of the World's Forests 1997* (Rome: 1997).

Ecosystem Conversion Spreads

Janet N. Abramovitz

Around the world, the conversion, degradation, fragmentation, and simplification of ecosystems has been extensive—all negative trends that are accelerating. (See Table 1.) In many countries, including some of the largest, more than half of the original territory has been converted from natural habitat to other uses, much of it unsustainably and irreversibly.[1] Even countries that remained relatively undisturbed until the 1980s have lost significant portions of remaining intact ecosystems in the last decade. Lost with these natural ecosystems are the valuable goods and services they provide.

Examining the area converted is one way to approximate the footprint of human activity on the planet. "Converted area" is defined here as natural ecosystems that have been transformed into cropland, permanent pasture for livestock, forest plantations, human settlements, or land for other activities. It represents the best estimate of historic conversions. A measure of more recent activity is "natural forest conversion" during the 1980s.

The data show that many large countries (those of more than 100 million hectares)—such as Argentina, Australia, Ethiopia, India, Mexico, Sudan, and South Africa—have had more than half of their land area converted.[2] Thus a large share of the Earth has been converted. Many large countries also have a high recent forest conversion rate. By virtue of their size, some countries are so large that even a "moderate" amount of historic conversion (25–49 percent) or recent forest conversion translates into large areas of ecosystems altered. Examples of such countries are Brazil, Bolivia, Colombia, Indonesia, Iran, the United States, Venezuela, and Zaire.[3]

Many countries whose ecosystems have remained relatively intact until recently began to undergo rapid change in the 1980s, such as the nations of Southeast Asia and South America, losing 10–30 percent of their forests in a single decade.[4] In some places, like Cambodia, Laos, and Suriname, the pace is accelerating and all remaining intact ecosystems may be converted within the next few years.[5]

Inland and coastal wetlands are among the most highly altered ecosystems worldwide. Europe has lost half to two thirds of its wetlands; the continental United States has lost 53 percent; and Asia, 27 percent.[6] Developing nations have also experienced extensive wetland conversions. The Philippines has lost 78 percent of its inland and coastal wetlands and mangroves.[7] Ecuador has lost 70 percent of its coastal habitat, and Thailand and Indonesia, about one third.[8]

Human activity routinely converts underwater habitat as well. One common fishing method—bottom trawling—may be the most widespread yet least known force of underwater habitat conversion. The area affected is extensive. According to Elliot Norse, president of the Marine Conservation Biology Institute, many of the world's fishing grounds are completely plowed over one to three times a year, and some, more than that.[9] He has estimated, for example, that "a small fleet of 100 shrimp boats...[trawls]...25,000 km^2 each year, an area larger than the state of Massachusetts. The U.S. alone has thousands of bottom trawlers."[10] Much the way logging transforms a complex forest ecosystem into a simpler system that supports far fewer species, trawling transforms the underwater habitat.

There are some caveats and exceptions to these data. First, it is difficult to estimate forest quantity and quality. Forests, for example, can be severely degraded and still meet the Food and Agriculture Organization's definition of "forest."[11] Thus forest conversion rates very likely underestimate the true extent of deforestation. Second, historic conversion may be hard to determine in areas that have had intense human settlement for millennia. And third, for many places data are simply not available.

The final, and perhaps most important, caveat is that absolute area alone cannot tell the whole story. Countrywide statistics can mask serious changes at a subnational or ecosystem level. In Egypt, for example, while only 8 percent of its total area has been converted, all of the Nile river system—by far the most productive part of the country—has been highly modified.[12] Temperate rainforests cover only a small fraction of the nations of the western hemisphere, yet they are one of the most threatened ecosystems—a fact not apparent in national data.

Furthermore, area statistics cannot measure the lost goods and services of a habitat, nor reflect its relative importance or uniqueness. Nonetheless, a conservative yardstick such as area converted does provide a crude picture of human impact on the planet.

TABLE 1: HISTORIC EXTENT OF ECOSYSTEM CONVERSION AND RECENT RATE OF NATURAL FOREST CONVERSION IN SELECTED COUNTRIES

COUNTRY	LAND AREA (million hectares)	SHARE OF TOTAL AREA CONVERTED[1] (percent)	SHARE OF NATURAL FORESTS CONVERTED, 1980s (percent)
Africa			
Sudan	237.1	52	10
Zaire	226.7	10	6
South Africa	122.1	78	8
Nigeria	91.1	80	7
Tanzania	88.4	44	12
Mozambique	78.4	60	7
Zambia	74.3	48	10
Madagascar	58.2	47	8
Côte d'Ivoire	31.8	53	10
The Americas			
United States	957.3	45	n.a.
Brazil	845.7	28	6
Argentina	273.7	62	6
Mexico	190.9	52	12
Bolivia	108.4	27	11
Colombia	103.9	44	6
Venezuela	88.2	25	12
Paraguay	39.7	60	24
Ecuador	27.7	19	17
Asia/Oceania			
Australia	764.4	60	n.a.
India	297.3	65	6
Indonesia	181.2	27	10
Thailand	51.1	43	29
Malaysia	39.9	15	18
Viet Nam	32.5	26	14
Philippines	29.8	36	29
Cambodia	17.7	25	10
Bangladesh	13.0	81	33
Europe			
Russian Federation	1,699.6	12	n.a.
Spain	49.9	60	n.a.

[1]Converted land represents forest plantations, croplands, and permanent pastures except for Australia, the Russian Federation, Spain, and the United States, which exclude forest plantations.
SOURCES: U.N Food and Agriculture Organization (FAO), "FAOSTAT-PC," electronic database, Rome, 1995; FAO, *Forest Resources Assessment, 1990: Global Synthesis* (Rome: 1995); World Resources Institute et al., *World Resources 1996–97* (New York: Oxford University Press, 1996).

Primate Diversity Dwindling Worldwide

John Tuxill

ollectively, the 233 known species of primates on Earth are in trouble. A recent report by the World Conservation Union–IUCN estimates that nearly half of all apes, monkeys, lemurs, and lorises are threatened with extinction (see Table 1)—a higher proportion than for any other major group of mammals, and higher also than for birds.[1] An additional 18 percent of all primate species are in decline, and may soon reach threatened status.[2]

There is, of course, one glaring anomaly in this trend in primates: the spectacular rise this century in the numbers and ecological demands of the human species. While precise numbers are not available, human beings appear to far outnumber all other primate species combined.[3] Unfortunately, in our treatment of Earth's ecological communities and biotic resources, we are proving poor companions for our closest living relatives.

Although nonhuman primates inhabit a variety of environments, from the snowy slopes of Chinese mountains to the searing deserts of Saudi Arabia and Yemen, their stronghold is the belt of tropical forest that girds the Earth's equator.[4] This emphasis, combined with primates' relatively well monitored status, makes them particularly good bellwethers of the ecological health of tropical forests.

Brazil is home to the most primate species—77 at present, with new species discovered in each of the past six years.[5] Indonesia (with 33 species), Zaire (33 species), and Madagascar (30 species) are additional centers of primate diversity.[6] Madagascar's nonhuman primate fauna is exceptional because it consists entirely of lemurs, the most ancient primate lineage alive today. Moreover, lemurs are endemic to Madagascar and the nearby Comoros Islands—that is, found nowhere else in the world.[7]

Humankind's greatest impact on other primates is our tendency to alter their habitat for our own use. Habitat loss, particularly the fragmentation and conversion of tropical forests by road-building, agricultural expansion, and other activities, contributes to the decline of nearly 90 percent of the 96 currently threatened primate species.[8]

Along Brazil's Atlantic coast, the rainforest that once covered 1 million square kilometers has, over three centuries, been reduced to a mere 1–5 percent of its original extent.[9] Eleven of the twelve primate species endemic

TABLE 1: SHARE OF PRIMATE SPECIES THREATENED, BY REGION OR COUNTRY

REGION/ COUNTRY	IN DANGER OF EXTINCTION[2]	VULNERABLE TO EXTINCTION	NEARING VULNERABLE STATUS
		STATUS[1]	
		(percent)	
Asia	31	39	20
Africa[3]	10	10	39
Madagascar	27	37	3
Latin America	14	19	n.a.[4]
All Regions	20	26	18[5]

[1]Criteria for classifying species include total population size (small versus large), population trends (declining versus stable), species range (broad versus restricted), spatial arrangement of populations (continuous versus fragmented), and degree of exploitation pressure on the species. [2]Combines two IUCN categories, "endangered" and "critically endangered." [3]Excluding Madagascar. [4]This category was not assessed for Latin American primates. [5]Likely an underestimate, since Latin American primates were not assessed.
SOURCE: World Conservation Union–IUCN, *1996 IUCN Red List of Threatened Animals* (Gland, Switzerland: 1996).

to the Atlantic forest, such as the wooly spider monkey and the golden lion tamarin, are threatened with extinction and now find shelter only in isolated, remnant forest patches.[10]

Loss of primary forest has also been severe throughout Asia, West Africa, and Madagascar. In Indonesia and Malaysia, orangutans—the arboreal great apes of Asia—have lost 80 percent of their forest habitat in the last 20 years.[11]

With the exception of chimpanzees, humans are the only primates that regularly and deliberately hunt other primates, and we take a heavy toll. At least 36 percent of all threatened primate species now face pressure from excessive hunting.[12] People most commonly hunt other primates for food. This occurs in all tropical regions, but is most detrimental where commercial demand for wild game or "bush meat" is strong.

In central African countries, primates constitute as much as 25 percent of the bush meat sold in markets.[13] In parts of the region, the sale of bush meat to traders supplying urban areas is the main income-generating activity available to rural families.[14]

Studies of selective logging operations in the Republic of Congo found that while logging itself did not harm the forests' ecological value for wildlife, hunters followed the network of logging roads and rapidly depleted local populations of primates and other large forest mammals.[15] Indeed, Congolese logging company employees supplemented their income by supplying hunters with weapons, ammunition, and transport in exchange for a share of the meat.[16]

We also capture other primates for use as pets and biomedical research subjects. The primate pet trade is particularly deleterious in Southeast Asia. Infant primates are preferred, and their capture invariably involves the death of their parent.[17] Although most nations have laws to regulate the hunting, sale, and possession of primates, these laws are unevenly enforced.

Most biomedical research involving primates takes place in industrial nations—the United States, United Kingdom, and Japan are the top primate-importing countries.[18] Approximately 40,000 primates are required each year for biomedical research worldwide.[19]

Fortunately, since the 1970s most countries with wild primate populations have restricted their export, and the demand for primates in research labs is increasingly met through captive breeding.[20] A major improvement occurred in 1994 when export bans on wild primates were adopted by Indonesia and the Philippines, who in previous years supplied 50–80 percent of all internationally traded primates.[21]

Despite these many problems, not all relations between humans and other primates are exploitative. People in many regions protect primates from harm by according them sacred status or making them taboo to hunt or eat. One of the rarest African monkeys, Sclater's guenon, survives at three sites in Nigeria in part because local residents treat the guenon as a sacred animal, and do not disturb it.[22]

Other endangered primates serve as "flagship" species, attracting public attention and generating protection for habitats or sites that otherwise might go unnoticed. In southeast Brazil, a program to conserve the golden lion tamarin has begun to restore degraded rainforest habitat.[23]

The famed mountain gorillas of Rwanda catalyzed the protection of an entire montane forest ecosystem, in part by earning Rwanda up to $10 million annually in foreign exchange through gorilla-watching tours.[24] Even after tourists and conservationists fled the country during the 1994 civil war, local park guards remained on duty and continued to patrol the gorillas' habitat, risking their lives without pay to protect the great apes.[25]

As these examples illustrate, we can follow a more humane code of conduct with our closest living relatives. Yet no matter how fully we appreciate, respect, and even venerate other primates, humankind ultimately must protect wild habitats and improve the way we use natural resources to give primates—as well as all other species—the room they need to survive.

Ozone Response Accelerates

Hilary F. French

In September 1997, the world will mark the tenth anniversary of a landmark in international environmental diplomacy: the Montreal Protocol on Substances That Deplete the Ozone Layer. This comes at a time when industrial countries have mostly finished the job of phasing out ozone-depleting substances (ODSs), while developing countries are just gearing up for the challenge. (See Table 1.)

The protocol mandated far-reaching restrictions in the use of certain chlorine- and bromine-containing chemicals that damage the ozone layer—the thin, vital veil in the stratosphere 10–50 kilometers above the Earth's surface that protects us from harmful ultraviolet (UV) radiation.[1] Scientists projected that without international action to reduce use of the offending chemicals, excessive radiation would have grave consequences for human and ecological health—including millions of additional skin cancer cases worldwide, sharply diminished agricultural yields, and extensive damage to aquatic life.[2]

In the 10 years since Montreal, rapidly advancing science and technology have contributed to a steady expansion in the requirements. The protocol has been significantly updated three times, now requiring that chlorofluorocarbons (CFCs), the most voluminous ozone-depleter, be phased out for nearly all uses in industrial countries by 1996 and in developing countries by 2010.[3] Use of several other ODSs is also restricted, including carbon tetrachloride, hydrochlorofluorocarbons (HCFCs), halons, methyl bromide, and methyl chloroform.[4]

At the time of Montreal, CFCs were widely used as propellants in aerosol cans, as foam-blowing agents, as solvents, and as coolants for refrigerators and air conditioners.[5] Yet by 1995, global production of CFCs was down 76 percent from its peak in 1988.[6] Global production of halons, carbon tetrachloride, and methyl chloroform has also declined steeply, although that of HCFCs and methyl bromide continues to climb.[7]

It was a notable achievement that the January 1996 deadline for phasing out production of CFCs for most uses in the industrial world passed reasonably smoothly in most places.[8] One exception was in the former Soviet Union and in some parts of Eastern Europe, where economic and political chaos has slowed progress in eliminating CFCs.[9] As a result, Belarus, Bulgaria, Poland, Russia, and Ukraine all had to ask for an extension on the deadline for eliminating CFCs and other ozone-depleting substances.[10]

The problems in Russia and elsewhere in the region are also contributing to another trouble spot in the implementation of the accord: the growth of a black market in smuggled CFCs that threatens to undermine significantly the phaseout in industrial countries. Russia is believed to be a major source of this trade.[11] Other countries on the suspect list include China and India.[12]

Meanwhile, industrial countries still face considerable challenges in phasing out methyl bromide and HCFCs. The protocol now requires them to reduce methyl bromide use gradually, and to phase it out altogether by 2010.[13] For HCFCs, a gradual reduction is also required, with a phaseout for all uses except the servicing of existing equipment by 2020.[14]

With the CFC phaseout mostly complete in industrial countries, attention is turning to the developing world. Developing countries are required to freeze CFC consumption levels in 1999, and to phase out use of these chemicals altogether by 2010.[15] HCFC use is to be frozen after 2015 and ended by 2040. No phaseout has yet been agreed for methyl bromide.[16]

On the encouraging side, 58 developing countries have stated a commitment to phase CFCs out earlier than required, in some cases virtually on the same schedule as industrial countries.[17] Indeed, consumption of ODSs is reportedly already falling in a number of developing countries—including Argentina, Brazil, Colombia, Egypt, Ghana, and Venezuela.[18]

On the other hand, consumption continues to rise rapidly in some countries with large populations, including China, India, and the Philippines.[19] Altogether, developing coun-

TABLE 1: USE OF CFCs AND HALONS IN SELECTED COUNTRIES, 1986 AND 1994[1]

| COUNTRY | USE | | CHANGE |
	1986	1994	
	(tons weighted by substance's ozone-depleting potential[2])		(percent[3])
China	46,600	90,900	+ 95
European Community	343,000	39,700	− 88
Russia	129,000	32,600[4]	− 75
Japan	135,000	19,700	− 85
South Korea	11,500	13,100	+ 15
Mexico	8,930	10,800	+ 21
Brazil	11,300	7,780	− 31
Thailand	4,660	7,230	+ 55
India	2,390	7,000	+193
Argentina	5,500	4,950	− 10
Canada	23,200	4,850	− 79
Malaysia	3,840	4,760	+ 24
Philippines	1,920	4,010	+109
Australia	18,600	3,890	− 79
Venezuela	4,590	3,130	− 32
Indonesia	1,710	2,880	+ 69
South Africa	18,700	2,420	− 87
Poland	10,600	1,680	− 84
Ukraine	1,850	1,530	− 17
United States	364,000	−91	− 100

[1]"Use" is production (the amount of a substance produced in a year plus stock at the end of the year minus stock at the beginning of the year) plus imports minus exports minus feedstock use. Thus, if a lot of stockpiled material is used as feedstock to make other chemicals, the consumption number can be negative, as it is for the United States. [2]Compounds vary in their ability to deplete ozone. These numbers reflect the tonnage of the various CFCs and halons in Annex A of the Montreal Protocol (CFC-11, CFC-12, CFC-113, CFC-114, CFC-115, Halon-1211, Halon-1301, and Halon-2402) multiplied by their respective ozone-depleting potentials (ODPs). The ODP value is the ratio of a given compound's ability to deplete ozone compared with the ability of a similar mass of CFC-11. [3]Percentages may differ from the data due to rounding. [4]Data are for 1993.
SOURCE: United Nations Environment Programme, "The Reporting of Data by the Parties to the Montreal Protocol on Substances that Deplete the Ozone Layer" (Nairobi: 12 September 1996).

tries' consumption of CFCs and halons increased by about one third from 1986 to 1994, albeit from a small base.[20]

Unfortunately, the overall global decline in production of ozone-depleting substances has not yet translated into a healing ozone layer, as it takes years for CFCs and other ozone-depleting compounds to reach the stratosphere, and some last for centuries once there.[21] Current estimates suggest that if all countries comply with the November 1992 amendments to the Montreal Protocol, chlorine and bromine concentrations in the stratosphere will peak toward the end of this decade.[22] The ozone layer will then begin to mend gradually, though a full recovery is not expected until about 2050.[23]

The world is thus currently suffering through the period in which the ozone layer will likely be most severely damaged. Some of the largest "ozone holes" on record have been experienced above the Antarctic over the last few years as a result, and ozone losses—along with levels of UV radiation—have increased over populated and agriculturally abundant corners of the Earth as well.[24]

Given these sobering realities, it is far too early to declare victory on the ozone front. Though the world community can be heartened by its ozone accomplishments to date, continued vigilance is required to ensure that the ozone experience remains a success story.

Harmful Subsidies Widespread

David Malin Roodman

Governments around the world are spending at least $500 billion of taxpayers' and consumers' money each year in support of activities that harm the environment, such as mining, logging, irrigation, pesticide and fertilizer use, crop and livestock production, energy consumption, and driving.[1] (See Table 1.) Though important in modern societies, these activities also deplete resources and pollute; as a result, subsidizing them makes economies less environmentally sustainable.

Subsidies for logging, for instance, are accelerating forest degradation.[2] Those for coal production are directly adding to local problems such as land disturbance and water pollution while contributing on a global scale to atmospheric concentrations of heat-trapping carbon dioxide.[3] High agricultural payments in industrial countries have been found to correlate with higher rates of fertilizer use,

thus increasing water pollution and soil degradation.[4]

The oldest and perhaps most environmentally destructive subsidies in western industrial countries are the ones that artificially stimulate resource-intensive industries on public lands in the name of economic growth. They date to a time when nations settled by European emigrants, including Australia, Canada, and the United States, were trying to encourage settlement of newly claimed lands by giving away resources.[5] But with those frontier days over, the special supports given to mining, logging, and livestock raising are ever harder to justify.

Hardrock mining, for example, is essentially free on public lands in Canada and the United States.[6] Since 1873, the U.S. government has given away roughly $242 billion in gold, silver, and other minerals—equal to

TABLE 1: SELECTED SUBSIDIES FOR ACTIVITIES THAT HAVE HARMFUL SIDE EFFECTS

Activity	Examples of Subsidies
Mineral Production	Low or zero royalties on oil and other minerals; aid for coal production in Germany, Russia, and other industrial countries.
Logging	Low timber royalties in developing countries; below-cost sales in North America and Australia.
Fishing	Billions of dollars per year in subsidies for fuel, equipment, and income support for fishers worldwide.
Agricultural Inputs	$13 billion a year lost on public irrigation projects in developing countries; billions more lost in industrial ones; subsidies for pesticides and fertilizers in some developing countries.
Crop and Livestock Production	$302 billion in annual support for farmers in western industrial countries; low fees for grazing on public lands in North America and Australia.
Energy Use	$111 billion in fossil fuel and power subsidies in developing countries each year; comparable losses in rest of world.
Driving	In excess of $111 billion a year in the United States in costs of roads, related services, and tax breaks over what drivers pay in fuel taxes and other fees.

SOURCE: Worldwatch Institute, based on sources cited in endnote 1.

$900 for every American alive today.[7] In Victoria, Australia, the provincial government spends some $170 million more building logging roads in public forests each year than it earns selling the wood to private loggers.[8] The U.S. government lost $300–400 million a year in the same fashion in the early 1990s.[9]

In the same vein, many developing countries have been cashing in their natural resources, usually for much less than they are worth, to jump-start economic growth. But the results generally have been slower growth and more poverty. The government of Indonesia, for example, has been selling rainforest logging rights since the 1960s far below market prices. In 1990, it collected $2 billion less in timber royalties than it could have—more than other countries gave it in development aid and loans.[10] Yet in some parts of the country, subsidized logging has destroyed more jobs than it has created, by cutting local people off from land on which they once grew cash crops.[11]

More generally, a recent Harvard University study found that on average, the more a developing country's economy depended on natural resource exports in 1971, the slower it grew in the 1970s and 1980s.[12] Encouraging logging, mining, and oil drilling has usually hurt, not helped, developing economies.

Numerous subsidies are meant not so much to speed economic change as to stop it: to protect domestic industries—including oil production, agriculture, fishing, and coal mining—for the sake of national security or jobs. Again, most do not work, at least at a reasonable cost.

In western Germany, for instance, the hard coal industry has become increasingly uncompetitive despite rising subsidies, which the government has offered in the name of protecting mining jobs. Partly as a result, employment has fallen 50 percent since 1982, and the subsidy per job protected reached

$72,800 in 1995.[13] It would be cheaper now to close the mines and pay miners a comfortable salary not to work. Similarly, fishing subsidies in many countries have contributed to overfishing, which depresses fish stocks and hurts the fishers meant to be helped.[14]

Many nations subsidize energy and water use to aid the poor. But, typically, the middle- and upper-income people who need the subsidies least receive them most because they buy the most energy and water. In Indonesia, across-the-board kerosene subsidies have cut the cost of living for the poorest fifth of the population, but 90 percent of the subsidies benefit better-off people.[15] Targeting the subsidies at the neediest recipients would give 10 times the benefit for the same public cost, or the same benefit for one tenth the cost.

Subtle subsidies also arise when governments fail to pass the full costs of infrastructure back to users. In the United States, governments raise some $111 billion less from tolls, vehicle registration fees, and fuel taxes each year than they spend on road construction, traffic management, and the income tax exemption for employer-provided free parking.[16] Those subsidies—worth 20¢ per liter of gasoline or diesel sold (78¢ per gallon)—encourage driving by making it appear cheaper than it is, contributing to pollution, oil dependence, and traffic jams that chew up billions of hours of people's time.[17]

In sum, most subsidies for environmental destruction can actually be eliminated or substantially trimmed in ways that do not hurt their effectiveness, or that even improve it, while saving money and helping the environment. With the global tax burden standing at roughly $7.5 trillion a year, the savings from such reforms could effectively reduce taxes by at least 7 percent, which would encourage work and investment even as it reduced the economic damage caused by subsidies.[18]

Economic Features

Sustainable Development Aid Threatened   Hilary F. French

At the June 1992 Earth Summit in Rio de Janeiro, a great deal of attention was focused on how to generate sufficient international funds to help developing countries implement the commitments made there.[1] Five years later, little of the promised money has been delivered.

Instead, overall aid levels have declined rather than risen in real terms since Rio. According to the Organisation for Economic Co-operation and Development, development assistance expenditures totalled $58.9 billion in 1995.[2] Though this was just slightly below the 1994 figure of $59.2 billion, in real terms it represented a decline of some 9 percent.[3] The most substantial decline was from the United States, which reduced its aid contribution by a quarter—from $9.9 billion in 1994 to $7.4 billion in 1995.[4]

The Agenda 21 action plan produced by the Rio conference estimated that developing countries would need some $125 billion in development assistance annually on top of hundreds of billions in domestic resources in order to put the plan into practice.[5] The conventions on biological diversity and on climate change signed at Rio also called for developing countries to receive "new and additional" funding, though neither contained any precise figures or commitments.[6]

It was widely recognized at the time of Rio that $125 billion was an ambitious target. Agenda 21 noted that the figure was "indicative" only, and reflected "order-of-magnitude" estimates.[7] Indeed, it was more than twice the level of total development assistance at the time, which was then just under $60 billion.[8] Yet Agenda 21

included a reaffirmation by donor countries of their earlier pledge to the United Nations to increase aid levels to 0.7 percent of their gross national products (GNP).[9] If all countries carried through on this pledge, total aid levels would approach $125 billion.[10]

Until the early 1990s, the United States had for decades been the world's largest aid donor.[11] Yet in 1995 it sank to fourth place, behind Japan, France, and Germany.[12] (See Table 1.) Delays in budget appropriations by the U.S. Congress were partly but not entirely to blame.[13] Even without them, U.S. aid contributions would have declined by almost 10 percent.[14] Altogether, in 1995 donors contributed just 0.27 percent of their collective GNP in aid—the lowest ratio since the 0.7 target was set in 1970.[15]

These declining overall aid levels have contributed to an acute shortage of funds for multilateral environmental and sustainable development initiatives. For instance, Agenda 21 called for the creation of an Earth

TABLE 1: DEVELOPMENT ASSISTANCE CONTRIBUTIONS, TOP 15 COUNTRIES AND TOTAL, 1995

COUNTRY	AS SHARE OF GNP (percent)	TOTAL (million dollars)
Japan	0.28	14,489
France	0.55	8,443
Germany	0.31	7,524
United States	0.10	7,367
Netherlands	0.81	3,226
United Kingdom	0.28	3,157
Canada	0.38	2,067
Sweden	0.77	1,704
Denmark	0.96	1,623
Italy	0.15	1,623
Spain	0.24	1,348
Norway	0.87	1,244
Australia	0.36	1,194
Switzerland	0.34	1,084
Belgium	0.38	1,034
Total, All Countries	0.27	58,894

SOURCE: Organisation for Economic Co-operation and Development, *Development Co-operation*, 1996 Report of the Development Assistance Committee (Paris: 1997).

Increment as part of the tenth replenishment of the World Bank's soft-loan arm for the world's poorest countries, the International Development Association (IDA).[16] Not only did the Earth Increment fail to materialize, but the overall IDA program has been short-changed as a result of the failure of the United States—traditionally the single largest donor—to come through with promised funds.[17]

The Global Environment Facility (GEF) faces similar challenges. Governments created this new fund on an interim basis in 1991 under the joint management of the U.N. Development Programme (UNDP), the U.N. Environment Programme (UNEP), and the World Bank. Its mandate is to finance investments in developing countries that help preserve the global commons—specifically international waterways, the atmosphere, and biological diversity.[18] In March 1994, governments agreed to make the GEF permanent and to replenish it with $2 billion in new resources to be spent over the next three years.[19] Governments subsequently designated the GEF as the interim funding arm for both the biological diversity and the climate pacts.[20]

Yet, as with IDA, the GEF has suffered from a failure on the part of the United States, Italy, and a few other donors to deliver fully on their commitments.[21] This is casting a shadow over ongoing negotiations aimed at recapitalizing the GEF for another three years.[22] It has also resulted in stagnating spending levels—the $315 million in GEF projects allocated for 1996 represented a slight decline from the $322 million approved for 1992.[23]

Meanwhile, U.N. agencies that are central to the sustainable development struggle are finding themselves fighting for their budgets. Despite calls in Agenda 21 for major roles for both UNEP and UNDP, these two programs have seen their relatively meager budgets shrink in the years since Rio, even when measured in current dollars.[24] UNEP received $98 million in funding in 1996, down from $105 million in 1992.[25] As for UNDP, it spent some $1.38 billion in 1995,

down from $1.42 billion in 1992.[26]

One agency that is faring better is the U.N. Population Fund (UNFPA); its spending increased from $212 million in 1990 to $312 million by 1995.[27] Many donors were energized to spend more money on international population programs as a result of the 1994 U.N. Conference on Population and Development. Japan and several European donors increased their support substantially, although U.S. contributions have fallen off sharply.[28] Yet when measured in constant dollars, government contributions to UNFPA do not look nearly so impressive.[29]

When world leaders gather in New York in June 1997 for a special session of the U.N. General Assembly on progress in the five years since Rio, there is bound to be considerable discussion of the failure to raise sufficient international funding for sustainable development. This failure must be reversed if the world is to have a fighting chance of turning Rio's noble rhetoric into on-the-ground results.

Food Aid Falls Sharply

Gary Gardner

International food aid fell in 1995, continuing a sharp decline that began in 1992. After peaking at 15.1 million tons that year, the grain component of food aid—which accounts for more than 90 percent of global food assistance—fell by more than half by 1995, to 7.2 million tons.[1] (See Figure 1.) The depressed levels of aid are well below the 10 million tons of annual giving promised by the international community more than 20 years ago, at the World Food Summit in 1974.[2] And the cuts come at a time of projected increases in the need for food assistance.[3]

Food aid includes emergency food shipments, concessional sales of food, and food donated as a component of development projects.[4] The bulk of food aid cuts has come in concessional sales, which fell from 75 percent of total food aid in the 1960s and 1970s to 43 percent in 1994.[5] Emergency food aid has been least affected by the cuts, and now assumes a growing share of the shrinking food aid pie—some 30 percent in the 1990s compared with 10 percent in the 1970s.[6]

Most of the world's food aid comes from the United States, the European Union, Canada, and Australia.[7] The United States has long been the leading donor, but its share of total food aid fell from 63 percent in 1993 to 41 percent in 1995.[8] Donations from this country and from the European Union are now roughly equal.[9] (Japan, a net food importer with little capacity to donate food, offers financial assistance to low-income nations for food purchases.)[10]

Africa claimed 40 percent of total grain food aid between 1991 and 1995, while a quarter of the total went to Asia.[11] Latin America and the Near East each received 11 percent on average.[12] In the 60 countries that received food aid between 1984 and 1994, the assistance accounted for some 5 percent of grain consumption. But dependence varied

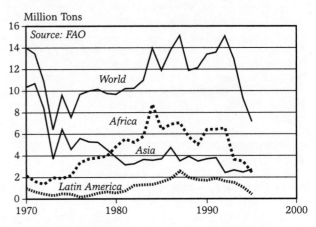

Figure 1: Grain Food Aid, Total and by Region, 1970–95

greatly by country.[13] In 11 Latin American countries, for example, food aid covered between 18 and 31 percent of food grain consumption.[14]

Cuts in grain food aid are occurring where the aid is needed most. The share of aid directed to the world's most vulnerable countries—those that are not self-sufficient in food and too poor to import all they need—has fallen from 90 percent in the 1980s to between 67 and 88 percent in the 1990s.[15] These "low-income, food-deficit" countries were hit especially hard when grain prices skyrocketed in 1995 and 1996. Their import bills jumped by 35 percent over 1994 levels, as food aid's share of grain imports fell to only 8 percent, compared with 20 percent in the mid-1980s.[16]

The withering of food aid has many roots. To some degree, cutbacks are good news: giving in peak years was propelled in part by disastrous production in some African countries (in the late 1980s) and in the former Soviet Union (in the early 1990s).[17] As these crises passed, the need for food assistance was reduced. But less benign forces are at work as well.

The end of the cold war removed the overarching rationale for foreign assistance by the world's superpowers, who saw aid (including food aid) principally as a tool of foreign policy.[18] From the early 1950s until 1992, development aid expanded continuously. But in 1992 levels of development aid began to drop, and emergency assistance assumed a greater share of the total.[19]

Cold-war food aid programs also had the happy advantage of eliminating unwanted surpluses in donor countries. But as subsidies to agriculture are reduced or eliminated in donor countries, and as global food demand rises, "mountains" of surplus grain may no longer be the norm.[20] Indeed, in 1996 grain reserves fell to their lowest level ever (measured in days of consumption) after being drawn down in seven of the previous nine years.[21] Tight supplies of grain have historically led to cuts in food aid, even though aid is only a tiny share of total grain production—typically from less than half a percent to nearly 1 percent.[22]

Skepticism about the effectiveness of food aid programs is another reason for cuts in giving. Some critics charge that foreign aid creates a culture of dependency, and point to Chile and Taiwan as cases where economic growth took off after aid was discontinued.[23] By increasing domestic supplies, they argue, food aid depresses domestic food prices, which reduces farmers' incomes and lessens their incentive to produce. Other critics assert that aid does not address the root causes of poverty, but simply props up unjust economic structures.[24] Advocates of aid counter that the problem is not assistance itself, but the way it is administered, noting that governments often misuse aid, and that nongovernmental organizations are the proper entities for handling it.[25]

Finally, some donors cite budgetary pressures to justify the cutback in giving. This was the rationale given by the United States in March 1995 when it notified the Food Aid Committee of the International Grains Council that it would cut its annual donor commitment from 4.47 million tons to 2.5 million tons.[26] As government budgets in donor nations tighten, besieged foreign assistance

programs are especially vulnerable to cuts.

In contrast to the trend in giving, the need for food assistance is expected to rise sharply over the coming decade. Using optimistic assumptions about population growth, foreign-exchange earnings, and yield increases in recipient countries, the U.S. Department of Agriculture (USDA) estimates that grain donations will need to rise to 27 million tons in 2005 just to maintain current levels of consumption and to meet emergency needs.[27] To boost consumption to minimum nutritional levels, some 30 million tons of food aid will be needed. This is more than four times the level of giving in 1995/96.[28]

Africa will continue to need the most food aid.[29] African requirements are expected to double by 2005, even under the most optimistic USDA scenario, and will account for 55 percent of global need.[30] By 2005, Africa alone is projected to need more than the entire global supply of food aid.[31]

R&D Spending Levels Off

Michael Renner

It is widely accepted in modern societies that investing in research and development (R&D) programs in such varied areas as product innovation, productivity and competitiveness, health care, food grain yields, protection of the environment, and national defense is essential to the welfare of society.[1]

The members of the Paris-based Organisation for Economic Co-operation and Development (OECD, grouping primarily rich Western countries) account for the vast majority of global R&D expenditures. Following strong growth in earlier decades, the R&D expenditures of these governments have stagnated at roughly $398 billion annually since 1991.[2] (All data are in 1995 dollars.) This levelling off is due to economic recession and tight governmental budgets.[3] U.S. federal government support for nonmilitary R&D may suffer cuts if plans to eliminate the budget deficit by the year 2002 are implemented.[4] The Japanese government, in contrast, is planning to boost its traditionally meager R&D funding by about 50 percent during the remaining years of this decade.[5]

On average, OECD countries are spending slightly more than 2 percent of their gross domestic product on R&D, or about $400 per person.[6] The United States, Japan, Germany, France, Britain, Italy, and Canada account for 90 percent of OECD spending, with the United States alone responsible for 44 percent.[7] Indeed, U.S. expenditures are so large that they surpass those of the next four largest spenders combined.

Most developing countries remain marginal players, even though a few of them have substantially stepped up investments.[8] For instance, South Korea spent about $10 billion in 1994, and Taiwan, $4.5 billion.[9]

Meanwhile, the severe shock of economic adjustment in formerly Communist countries has been felt in their R&D programs as well. Russian R&D expenditure fell by 75 percent in 1990–93, to about $6.4 billion.[10]

Among OECD countries, governments on average are contributing a little more than one third of all R&D funds; industry adds close to 60 percent, with the remainder covered by universities and other nongovernmental groups.[11] Businesses carry out most R&D projects (absorbing two thirds of all funds), followed by universities (17 percent), government agencies (13 percent), and nonprofit organizations (3 percent).[12] Aerospace, electronics, pharmaceuticals, office machines, and computers are among the industries attracting most private R&D investments.[13]

Although private business plays a dominant role, public R&D investments are nevertheless important. (See Table 1.) This is particularly the case in supporting the basic research that yields important scientific discoveries but is often years or decades removed from accruing commercial benefit. But it is also true in applied research, where new technologies must compete with well-established ones. Government R&D programs are crucial when businesses fail to invest due to deceptive market signals; one example is R&D in energy efficiency and solar technologies in periods of low oil prices.

One area in which a large governmental role has long been accepted is military R&D. With the end of the cold war, these global expenditures have fallen by 41 percent, from a peak of $124 billion in 1988 to about $73 billion in 1993.[14] In OECD countries, one third of all government R&D spending goes to military programs.[15] Yet this average figure masks enormous differences among members: 55 percent in the United States compared with a mere 6 percent in Japan.[16]

At a comparatively small $10 billion annually, energy projects account for just 6 percent of total OECD government R&D.[17] Responding to the oil crises of the 1970s, OECD governments more than doubled their investments in energy R&D between 1974 and 1981 to $16 billion; but budgets declined just as quickly after oil prices fell in the early 1980s.[18] Japan, highly dependent on imported energy, is the only country to resist the downward trend; the government devotes 21 percent of its total R&D funds to energy, mostly to nuclear energy programs.[19]

Since 1974, OECD governments have invested a cumulative $247 billion in energy R&D.[20] The bulk of these funds went to nuclear programs—conventional reactors, breeders, and nuclear fusion; much smaller amounts have been devoted to renewable energy and energy efficiency.[21] Rising to roughly $1 billion in 1995, efficiency programs are now once again funded at the level they had briefly reached in 1980.[22] Renewable energy R&D, too, is growing, but remains at less than half its $2 billion peak in 1980.[23]

At about $12 billion annually, global agricultural R&D investment is roughly on a par with energy R&D. It has been stagnant in the 1990s, following rapid growth during the 1960s and 1970s and slower growth in the 1980s.[24] About half of global spending is taking place in developing countries, but their expenditures are equivalent to only 0.5 percent of their agricultural output, compared with 2–4 percent in industrial countries.[25] The U.N. Food and Agriculture Organization calls for at least a 50-percent increase in developing nations' funds to boost food production.[26]

R&D for environmental protection purposes is still a rather small component of overall R&D expenditures. It absorbs slightly more than 1 percent of OECD governments' spending, although the situation in individual countries varies widely.[27] The Japanese government, for instance, spends 0.5 percent; Germany, by contrast, devotes 3.7 percent.[28] The OECD's seven largest countries combined spend some $2.3 billion annually on such projects.[29]

The U.S. Environmental Protection Agency had a 1995 R&D budget of about $600 million.[30] But R&D funding under the broader "Environmental Sciences" category in the federal budget amounted to roughly $2.5 billion in 1995, including $970 million for atmospheric research.[31] And the U.S. National Science Foundation maintains that, more broadly defined, more than $6 billion was available for environment and natural resource–related R&D programs in the 1996 budgets of 12 different government agencies.[32]

TABLE 1: GOVERNMENT R&D BUDGET APPROPRIATIONS, SEVEN LARGEST WESTERN ECONOMIES, EARLY 1990S

CATEGORY	TOTAL (billion 1995 dollars)	SHARE (percent)
Military	62.6	36
Advancement of Knowledge[1]	39.2	23
Health	18.2	11
Civilian Space	15.5	9
Energy	10.7	6
Industrial Development	8.0	5
Agriculture, Forestry, and Fishing	5.6	3
Infrastructure[2]	3.8	2
Earth and Atmosphere	2.7	1
Social Development and Services	2.5	1
Environmental Protection	2.3	1
Total[3]	171.6	100

[1]Includes advancement of research and general university funds. [2]Includes transport, telecommunications, urban and rural planning. [3]Includes other categories not listed separately.
SOURCE: Based on National Science Board, *Science and Engineering Indicators 1996* (Washington, DC: U.S. Government Printing Office, 1996), Appendix Table 4-32.

Agriculture Grows in Cities   Toni Nelson

While today's cities may not seem well suited to producing food, the practice of urban agriculture provides an important source of food and income for many city dwellers.[1] (See Table 1.) Although the global contribution is difficult to assess because only a few local studies have been conducted, the most widely used estimate holds that approximately 200 million people currently participate in urban farming, providing 800 million people with at least some of their food.[2]

By 2025, two thirds of the world's population will live in cities, up from 45 percent in 1995.[3] Hunger is already a way of life in the cities of both rich and poor countries, and the growth of these urban areas will mean even more mouths to feed. In the United States, where more than three quarters of the population is urban, as many as 30 million people are hungry; in 1993, one person out of every six received some form of federal food assistance.[4]

In the developing world, where poverty affects a higher proportion of the population, even more people live near the edge of hunger. One person in five is undernourished in the Asia-Pacific region; in Africa, the figure is one in three.[5] In the cities of these regions, the urban poor spend the bulk of their incomes on food: more than half of all urban households in Egypt spend at least 60 percent of their total budget on food, for example, while in Ecuador 74 percent of urban households cannot afford even basic food purchases.[6]

For many of the people most at risk in cities, locally produced crops provide a crucial source of food. Approximately 20–30 percent of all household food supplies, particularly such vegetables as cassava leaves, amaranth, potatoes, and pumpkins, are grown in Dar es Salaam, Tanzania, producing some 100,000 tons of food annually.[7] In Singapore, urban farmers produce 25 percent of the city's vegetables and 80 percent of its poultry.[8] Cuba has recently begun to promote urban agriculture; employees of the Australian Conservation Foundation, which sponsors an urban gardening project in Havana, estimate that production there could expand from the current 5 percent of total consumption to 20 percent.[9]

Urban agriculture often provides income for urban families as well. In Dar es Salaam, farming employs one fifth of the population, and farmers there constitute the second largest occupation category, after small traders and laborers.[10] In Bogotá, a cooperative of more than 100 low-income women grows vegetables hydroponically on rooftops, earning the women up to three times more than their husbands make in semiskilled jobs.[11] Even households that produce food for their own consumption can save substantially: one study in Buenos Aires found that gardening can cut 10–30 percent off the cost of a family's food purchases.[12]

Farming in cities has another important benefit—it uses wastes that can otherwise contaminate local ecosystems and endanger people's health. According to the World Resources Institute in Washington, DC, an estimated 20–50 percent of solid waste goes uncollected in the cities of the developing world, and more than 90 percent of the sewage in those cities is discharged untreated directly into rivers, lakes, and coastal waters.[13] Industrial countries also struggle with the problem of waste management, as demonstrated by the islands of waste growing in Tokyo Bay as the city attempts to cope with the 22,000 tons of garbage it generates daily.[14]

Chinese cities provide perhaps the most dramatic example of how urban wastes can be used to grow food. Farmers in and surrounding 18 of China's largest cities produce 85 percent of the vegetables and more than half of the meat and poultry consumed there, while recycling most of the human wastes.[15] In 1991, for example, Shanghai collected 90 percent—about 8,000 tons—of the city's human waste each day, treated it, and sold it to local farmers.[16]

Raising fish on sewage is another profitable way to recycle human wastes and provide food at the same time. In Calcutta, a 12,000-hectare cluster of ponds produces more than 10 percent of the city's fish while processing 680,000 cubic meters of its wastewater daily.[17] Studies in Lima have found that fish ponds fed by partially treated wastewater could quickly recover their construction and maintenance costs through fish sales.[18]

Urban agriculture is not without its shortcomings, however. Improperly managed wastes can breed the bacteria that cause intestinal infections, and crops can absorb heavy metals and other contaminants from the soil or air. Contaminated wastewater used for irrigation in Santiago, Chile, was responsible for an outbreak of cholera in 1991. Fish grown in wastewater also may be contaminated and carry diseases such as hepatitis.[19] But many of these problems can be avoided through appropriate management practices.[20]

While most experts agree that urban farming is now both viable and desirable, they differ on its potential for feeding the world. Jac Smit, president of The Urban Agriculture Network, claims that urban farmers already produce 15 percent of the world's food, and that the practice has the potential "to produce one-quarter of the nutritional needs of the global population where they live."[21] Rachel Nugent, an economist who wrote the chapter on urban agriculture in the U.N. Food and Agriculture Organization's *1996 State of Food and Agriculture*, takes a more cautious view: urban agriculture requires much more planning than exists in the cities of most developing countries. "It is not a panacea to solve the most severe problems of food security in cities," she notes, and is "at best...a survival technique for the urban poor."[22]

TABLE 1: SELECTED EXAMPLES OF URBAN AGRICULTURE

CITY	AGRICULTURAL PRACTICES
Kampala, Uganda	More than one third of households grow crops on over half of the city's land area.
Havana, Cuba	More than 20,000 gardens supply about 5 percent of the city's food.
East Calcutta, India	150–300 tons of vegetables are produced daily on 800 hectares of old garbage dumps, providing jobs for about 20,000 people.
Hartford, Connecticut	The annual value of marketed production from urban farms ranges from $4 million to $10 million. Community and back-yard gardens produce an estimated additional $500,000 to $1.25 million worth of food a year.
Singapore	The city licenses just under 10,000 farmers on about 7,000 hectares. Urban farmers produce 25 percent of the city's vegetables and 80 percent of its poultry.
Berlin, Germany	Citizens cultivate 80,000 gardens on more than 3,200 hectares (4 percent of the city's total land area).

SOURCES: See endnote 1.

Gap in Income Distribution Widening   Hal Kane

In 1991, the richest 20 percent of the world's people earned 61 times more income than the poorest 20 percent, according to the United Nations Development Programme (UNDP).[1] Not only is this gap wide, it has been growing. Thirty years ago, the richest 20 percent received 30 times more than the poorest 20 percent—half as wide a gap as today.[2] Since then, the poorest one fifth saw their share of global income drop from 2.3 to 1.4 percent, while that of the richest group rose from 70 to 85 percent.[3]

As writer Robin Wright points out, "in the mid-1990s, the meek show no signs of inheriting the Earth."[4] The assets of the 358 billionaires in the world exceed the combined annual incomes of countries with the poorest 45 percent of the world's people.[5] From 1956 to 1980, 200 million people saw their per capita incomes fall, but from 1980 to 1993, more than 1 billion people had this experience.[6]

Distribution of income is by no means the only measure of equity or of well-being—other gauges include the distribution of land ownership and access to clean water, wood, education, health care, and many other resources and services. Brazil, for example, has the most unequal concentration of land in Latin America—the region with the worst land distribution; 45 percent of Brazilian land is owned by 1 percent of the population, an inequality that will have major bearing on that country's future.[7] Any comprehensive measure of Brazil would thus have to include land and other values as well as income. Nevertheless, income distribution remains a rough and approximate yet useful overall measure of access to resources and services.

Of the $23 trillion gross world product in 1993, $18 trillion was generated in industrial countries.[8] Only $5 trillion circulated in developing countries, where almost 80 percent of all people live.[9] The gap in per capita income between the industrial and developing worlds tripled over the past three decades—from $5,700 in 1960 to $15,400 in 1993.[10]

More than 3 billion people live in the 60 countries where incomes in the 1990s have been higher than ever before, most of them in Asia or among the countries in the Organisation for Economic Co-operation and Development.[11] The incomes of about 1.5 billion people in some 100 countries were lower in the 1990s than in earlier decades, according to UNDP.[12] Most of these people are in Africa, Latin America and the Caribbean, the Arab states, Eastern Europe, and the Commonwealth of Independent States. This means that economic growth has been failing over much of the past 15 years in about 100 countries, home to more than a quarter of the world's people.[13]

The United Nations has called this a spectacular economic advance for many and an unprecedented decline for others, resulting in remarkable inequalities.[14] (See Table 1.) The declines have far exceeded in duration, and sometimes in depth, those of the Great Depression of the 1930s in industrial countries.[15] The proportion of people with declining incomes more than tripled in the past 15 years, for example, from 5 percent to 18 percent.[16]

The economic transitions of the former Soviet territories and Eastern Europe have led to remarkably sudden income disparities. In Kyrgyzstan in early 1994, the income of the richest 10 percent of the population was 1.5 times as large as that of the poorest 10 percent—but by the end of that same year it was 10 times higher, according to UNDP.[17] In Russia, the nominal incomes of the richest 10 percent of households increased by 30 percent in 1994, while those of the poorest 10 percent rose by only 5 percent.[18] (Even though incomes rose, the cost of living in Russia went up much further, making most people relatively poorer.) These trends have been accompanied by rising crime rates and unemployment, growing poverty, worsening health, and a rising death rate.

In the United States, new Census Bureau research showed that from 1968 to 1994 the share of the nation's aggregate income that went to the top 20 percent of its households increased to 46.9 percent from 40.5 percent.[19]

During the same time, the share of income earned by the rest of the country's households either declined or remained stagnant.[20]

The average income for the top 20 percent of U.S. households grew to $105,945 in 1994 from $73,754 in 1968, a jump of 44 percent after being adjusted for inflation.[21] In contrast, the bottom 20 percent of households saw their income go up in constant dollars from $7,202 to $7,762, a 7-percent increase

during the same period.[22] In the United States, between 1975 and 1990 the richest 1 percent of the population increased its share of assets from 20 percent to 36 percent.[23]

Studies done in 1995 showed some of the importance of the distribution of income. Researchers from Harvard and Columbia concluded from statistical patterns in 70 countries that high income inequality creates social unrest and economic uncertainty that discourages investment.[24] Another statistical study concluded that more equal societies tend to have lower birth rates and to invest more in education.[25]

Yet the ways in which equality are achieved seem to be at least as important as the actual gap between rich and poor. East Asian countries that have had relatively equal distributions of wealth dating from before their industrialization have fared far better economically than countries such as Russia that levelled their differences by force.[26]

TABLE 1: INCOME DISTRIBUTION, SELECTED COUNTRIES, 1980S AND EARLY 1990S

COUNTRY	RATIO OF RICHEST ONE FIFTH TO POOREST
Industrial Countries	
Hungary	3.2
Japan	4.3
Germany	5.8
Norway	5.9
Israel	6.6
Canada	7.1
France	7.5
United States	8.9
Australia	9.6
United Kingdom	9.6
Russian Federation	11.4
Developing Countries	
India	4.7
Ethiopia	4.8
Indonesia	4.9
Viet Nam	5.6
South Korea	5.7
Ghana	6.3
China	6.5
Algeria	6.7
Thailand	8.3
Nigeria	9.6
Singapore	9.6
Zimbabwe	15.6
Chile	18.3
South Africa	19.2
Kyrgyzstan	22.8
Panama	29.9
Guatemala	30.0
Brazil	32.1

SOURCE: United Nations Development Programme, *Human Development Report 1996* (New York: Oxford University Press, 1996).

Electric Cars Hit the Road

Seth Dunn

There are approximately 7,500 electric cars in use today.[1] (See Table 1.) This estimate, based on data collected through the end of 1995, should be considered a lower boundary. The next several years are likely to witness a rise in the number of these vehicles as efforts to commercialize them accelerate and as reporting about their use improves.

Switzerland—with 1,200 electric cars—has the highest number per capita; 934 of these are very small, three-wheeled two-seaters.[2] The Swiss government has long been committed to solar and electric vehicles; it has promoted them through races, city demonstration projects, and tax breaks for purchasing electric cars.[3] It also held a referendum on clean air, has restrictions on nonelectric vehicles in villages, and hopes to have electric cars account for 8 percent of on-road vehicles by 2010.[4]

Germany has the largest number of electric cars overall, with a reported 2,452 on the road; very small passenger vehicles account for 1,250 of this total.[5] The German government offers tax exemptions for the purchase of electric cars, and several states subsidize purchases or lease these vehicles at rates equivalent to conventional ones.[6]

About 2,300 electric cars are found on U.S. roads.[7] Federal, state, and local governments offer a number of tax exemptions for buying them.[8] Most notably, the California Air Resources Board (CARB) has required that 10 percent of the vehicles sold by the seven biggest automakers in the state in 2003 be "zero-emission," which basically means they must be electric.[9] (In 1996, the state retreated from a stiffer requirement of 2 percent zero-emission by 1998.)[10]

The California mandate launched a race among large and small manufacturers to market electric cars. In December 1996, General Motors began leasing the first modern, mass-produced electric car in California and Arizona; Honda plans to make the first family-size model available for leasing in California in spring 1997.[11] Conventional cars converted to electric-powered have been

TABLE 1: ELECTRIC CARS, TOTAL AND BY COUNTRY, 1995

COUNTRY	TOTAL (number)
Germany	2,452
United States	2,306
Switzerland	1,200
France	468
Austria	346
Italy	296
Sweden	156
United Kingdom	93
Japan	83
Norway	60
Netherlands	21
Belgium	20
Canada	12
Finland	9
Total	7,522

SOURCE: Worldwatch estimate based on International Energy Agency, European Electric Road Vehicle Association, and Electric Vehicle Association of the Americas.

widely tested and available for several years in Europe, Japan, and the United States from several carmakers, including PSA Peugeot-Citroën, Renault, and Solectria.[12]

Consortia of government agencies, aerospace and electric companies, and environmental groups play a growing role in the electric car industry. Eight such organizations exist in the United States.[13] In addition, electric vehicle associations in North America, Europe, and Asia keep in touch and hold biannual meetings to share technologies and production plans.

Electric car production may end up being centered in the developing world, where labor costs are low and many other electronic products are made. To encourage this, Thailand—already producing electric versions of its three-wheeled "tuk-tuk" taxi—has combined tax exemptions for domestic electric car production with a tariff on gas-powered imports.[14] Korean carmakers Daewoo and Hyundai plan to produce lightweight electric cars in 1998.[15] China plans to supply raw materials and batteries for electric cars made in Taiwan; a domestic firm is working with PSA Peugeot-Citroën to develop a small electric car.[16]

Several obstacles stand in the way of wider use of electric cars. The purchasing costs of the first models remain above those of conventional cars, although their operating and maintenance costs may be significantly lower.[17] There has also been little effort to date to improve the energy storage of batteries, which are currently expensive to produce and provide a limited range between charges.[18]

The initial price of electric cars, however, is expected to drop sharply as manufacturers achieve economies of scale. Daniel Sperling of the University of California at Davis, using the price history of the conventional car as a basis for comparison, estimates the full-scale production of electric cars could pull prices to well below half their current level.[19] Tufts University's Global Development and Environment Institute projects dramatic price drops analogous to the electronics industry, with costs nearing those of conventional cars in the near future.[20] Some experts believe electric cars will compete directly with gasoline models within a decade.[21]

In parallel with cost reductions, battery improvements are expected to extend the range of electric cars significantly. Battery weight has already been cut by as much as 60 percent in the last decade, and further reductions are expected.[22] A technical panel appointed by the CARB has concluded that high-energy, high-power batteries could be in commercial production within five years, providing ranges above 150 kilometers and acceleration equal to internal combustion engines.[23] Other technologies being pursued for future electric cars include hybrid-electric drives—combining a gasoline engine and electric motor—flywheels, ultracapacitors, and fuel cells.[24]

Without waiting for these improvements in cost and range, low-cost, limited-use electric cars are growing in number. Many of these are available from small carmakers such as the Norwegian consortium PIVCO, whose City Bee is scheduled for sale later this year.[25] Large automakers have also entered the small-car niche: Mercedes-Benz, in a partnership with Swiss watchmaker Swatch, will offer a compact Smart car in 1998.[26]

Electric cars are also being integrated with public transport. The City Bee serves as a "station car" in a San Francisco Bay transit system program under which customers use credit cards to rent the vehicles for short distances.[27] A similar arrangement is being set up in the city and suburbs of Paris, where PSA Peugeot-Citroën and Renault are operating small electric cars for commuters.[28]

Several governments are encouraging this automotive transition. In addition to those already mentioned, Austria, France, Italy, Japan, Netherlands, Norway, Sweden, and the United Kingdom provide a mix of subsidies, tax credits, and exemptions for purchases.[29] Other forms of support include California-style mandates, government procurement of electric vehicles, demonstration projects and infrastructure, partnerships with industry, and research and development funding.[30]

Arms Production Falls   Michael Renner

Worldwide production and stockpiles of arms continued to decline in 1994 and 1995. (Although no more recent data are available, these trends appear to have continued in 1996 as well.) This trend is the product of a variety of arms control and peace agreements, and of economic difficulties experienced by several countries. The decline in the international trade of arms, however, came to an end in 1995. And numerical reductions are often offset by growing technological sophistication of weapons systems.

Global arsenals continue to be trimmed. The number of main battle tanks deployed worldwide, for instance, fell from roughly 172,000 to 119,000 between 1993 and 1996.[1] The number of combat aircraft declined from 40,000 to 31,000 in that same period; submarines, from 652 to 571; and major surface battleships, from 1,000 to 966.[2] There have also been substantial reductions in the numbers of artillery pieces, armored vehicles, attack helicopters, and other weapons systems.[3] Some of these weapons have been dismantled or destroyed; others, only put in storage.[4]

The Bonn International Center for Conversion (BICC) puts the combined reductions in five major weapons categories (tanks, artillery, combat aircraft, battleships, and submarines) since 1990 at 19 percent.[5] Weapons deployed by the armies of industrial countries fell by an average of 25 percent, with cutbacks reaching 56 percent in the former Soviet Union.[6]

Developing countries cut their arsenals less—by 12 percent on average.[7] Very large cuts took place in several countries where wars have ended: Angola, El Salvador, Ethiopia, Iraq, and Nicaragua. But several Asian countries, including Afghanistan, Bangladesh, Cambodia, Myanmar, South Korea, and Thailand, have dramatically bolstered their holdings.[8]

Budgets to procure new arms are declining in most parts of the world, and hence production has been falling as well. Data compiled annually by the Stockholm International Peace Research Institute (SIPRI) show that the military sales of the top 100 arms manufacturing companies in the world dropped from $186 billion in 1990 to $150 billion in 1994, the most recent data available.[9] (All data are in 1995 dollars.)

These numbers do not include companies in China and the former Soviet Union, which have had steep cuts as well. Russian arms producers have experienced by far the sharpest decline. According to one estimate, by 1995 arms production had plummeted to just one sixth of its 1991 level.[10]

Altogether, it appears unlikely that global military procurement spending now surpasses $200 billion a year.[11] With lower budgets for equipment purchases, employment in the arms industry is also falling. According to estimates by BICC, global employment in arms production declined from a peak of 17.6 million in 1987 to 11.1 million in 1995—a 37-percent reduction.[12] (See Figure 1.) The global outlook is for more job losses in coming years.

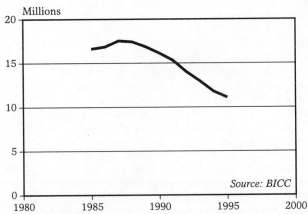

Figure 1: Global Arms Industry Employment, 1985–95

The largest cuts, in absolute numbers of people affected, have taken place in the former Soviet Union, China, and the United States.[13] In relative terms, though, Poland, Germany, the former Soviet Union, and South Africa had the most significant reductions among major producers—ranging from 46 to 62 percent since 1987.[14] Among smaller producers, Bulgaria, Hungary, and Iraq lost around 80 percent of their arms industry jobs; Romania, Spain, Belgium, Argentina, Iran, and Brazil lost roughly 60–70 percent.[15] In India, military industry employment has more or less held steady, and in Indonesia, North and South Korea, Syria, and Turkey, it is growing.[16]

A growing number of countries are producing at least a limited range of weapons domestically. The capacity to produce smaller-caliber arms is quite widespread. The 1993/94 edition of *Jane's Infantry Weapons* reported that some 1,700 different weapons are being produced by 252 manufacturers in 69 countries.[17] Still, international arms transfers play a very important role in equipping the armed forces of the great majority of countries, and are unavoidable for the more sophisticated types of military equipment.

According to the U.S. Arms Control and Disarmament Agency (ACDA), in 1995 world arms trade increased by 19 percent, to $31 billion, from the previous year.[18] However, since 1984, when arms worth a record $84 billion were transferred, the international arms trade has declined by 63 percent.[19]

In 1994, the last year for which regional data are available, developing countries accounted for 58 percent of all arms imports, but only for 7 percent of arms exports; a decade earlier, their shares were 76 percent and 10 percent, respectively.[20] ACDA reports that between 1983 and 1994, developing countries took delivery of 16,085 tanks; 28,368 armored personnel carriers; 27,969 artillery pieces; 900 naval combatant vessels and 37 submarines; 4,319 combat aircraft; 2,955 helicopters; and 44,315 surface-to-air missiles.[21] By the end of the 1983–94 period, deliveries in all these weapons categories had

fallen substantially.[22]

The ranks of exporters are far more concentrated than those of the importers. The top six suppliers account for 88 percent of total exports (with the United States alone responsible for 56 percent); on the import side, it takes the combined purchases of the top 44 recipients (primarily from the Middle East, East Asia, and Western Europe) to reach the same proportion of global imports.[23]

The declining sales of new weapons have been partially offset by transfers of used weapons, which have doubled in recent years.[24] This is because large stocks of armaments have become surplus in recent years and governments are looking for cheap ways to get rid of them. SIPRI estimates the global trade in secondhand arms at $28 billion during 1989–94.[25]

Close to 14,000 major used weapons systems (including ships, armored vehicles, artillery, and aircraft) were transferred, primarily by the United States and Germany, to other countries during 1989–94.[26] Used arms accounted for about one third of all systems sold.[27] Except for aircraft, used weapons transfers in 1994 exceeded sales of new ones in all other categories.[28]

Social
Features

Global Population Growing Older

Jennifer D. Mitchell

As a result of falling birth rates and rising life expectancy, the population's age structure is shifting. Decades of family planning have reduced the number of children born to women on average worldwide: between 1950 and 1996, births per woman dropped from an average of five to less than three.[1]

During this same period, antibiotics and vaccines, combined with better nutrition and sanitation, improved life expectancy in most countries. Average life expectancy increased by 21 years in developing countries and by 8 years in industrial ones.[2] As the number of seniors (people aged 60 and over) increases relative to the number of young people, the population is aging.

Some 540 million people around the world were 60 or older in 1995—almost twice as many as in 1965.[3] (See Figure 1.) Every day, another 30,000 people celebrate their sixtieth birthdays.[4] And given the swelling ranks of people between 20 and 50 years of age, this number is set to keep growing.[5]

In addition to the growing number, the proportion of seniors in the population is increasing. Currently, people over 59 represent just 9.5 percent of the total population; but by 2050, more than 20 percent of the world—one in every five individuals—will be in this age group.[6]

In industrial countries, on average, seniors already represent 18 percent of the total population, whereas in developing countries, the figure is 7 percent.[7] Yet developing countries have a larger absolute number of seniors: 330 million people over 59—almost two thirds of the global total—live there.[8] (See Figure 2.)

China's population is aging especially quickly. In just 27 years—between 2000 and 2027—seniors' share of the population there will jump from 10 to 20 percent.[9] This same

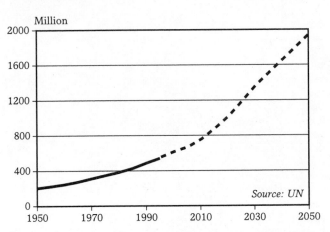

Figure 1: World Population 60 and Over, 1950–95, With Projections to 2050

increase took 60–100 years in some industrial countries.[10] The rapid shift is due to China's strict population policies, which lowered the fertility rate from 4.8 children per woman in the 1970s to 1.8 in 1996.[11] This, in combination with a 28-year increase in life expectancy since 1950, means that between 1995 and 2025, China's population over 60 will grow seven times as fast as its total population.[12]

Some of the "oldest" countries today are Sweden, Germany, Japan, and Italy, where more than 20 percent of the population is already over 59.[13] Remarkably, by 2025, half the population of Italy will be older than 50 and one third of the people in Japan are expected to be over 80.[14]

As the age structure shifts, populations are rapidly outgrowing the institutions intended to care for older individuals. In most industrial countries, social security schemes borrow from today's workers to pay today's retirees.[15] But as the number of retirees swells and the number of younger workers declines, the burden on the younger group increases. In industrial countries, five workers now support every retired person, but by 2030 this number will drop to just three.[16] In Japan the problem is even more severe: by

2025, the number of workers per retiree will drop from six to two.[17]

The retirement picture is further complicated by longer life expectancies and changing work patterns. When retirement was instituted in the United States, life expectancy was three years less than the retirement age; today, the average U.S. worker lives 11 years beyond the standard retirement age of 65.[18] Yet the share of men working in their late sixties is declining, further increasing the burden on other workers. In 1940, 52 percent of American men over 65 were working; by 1995, the figure was 16 percent.[19]

Most developing countries have virtually no social security or pensions, and people must rely on their families for support.[20] As birth rates decrease and families become smaller, the older generation leans on younger family members more heavily.[21] And often the children of elderly individuals are seniors as well; 65-year-old children are having to care for their 85-year-old parents. Urbanization, the erosion of traditional values, rapid development, and the dispersal of aunts, uncles, and children also make it harder for families to care for aging members.[22]

Aging populations change health patterns as well. The likelihood of suffering from chronic diseases such as cancer, heart disease, osteoporosis, or arthritis increases with age, raising overall medical costs.[23] Today, a 75-year-old in the United States uses 10 times as much medical care as a 40-year-old.[24] In developing countries, the health needs of seniors are mostly unmet.[25]

Other changing patterns are also apparent. In older populations, women outnumber men. They tend to outlive men by eight years in industrial countries; in some cases there are twice as many women over 59 as men.[26] In developing countries, the gap is narrower: women live only three years longer than men.[27] Older women typically earn less throughout their lifetimes, and are more likely to face extreme poverty in old age.[28]

As a growing number of people pass 60, more resources will be necessary to meet the needs of seniors. In countries where resources are scarce, money may have to be diverted from other areas, such as programs for the young, which could result in conflict between the generations.[29] Investment in areas such as education, however, benefit both old and young.

Education enables young individuals to accumulate resources and prepare for old age, and makes them more valuable to the community and the family when physical labor is no longer possible.[30] Education can also bridge cultural gaps between older and younger generations.[31]

Despite the potential strain on families, societies, and the economy, seniors play important roles in the family and community. In Chile, for example, 55 percent of seniors assist in the education of children and 51 percent teach vocational skills.[32] Helping families to care for their older members, and creating dynamic programs and policies that can be adjusted to reflect the changing population structure, will allow society to benefit from this growing human resource.

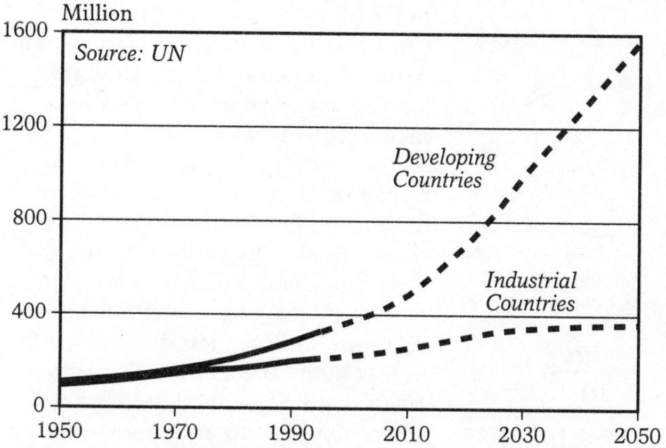

Figure 2: Population 60 and Over, Industrial and Developing Countries, 1950–95, With Projections to 2050

Noncommunicable Diseases Rising

Anne McGinn

In 1990, an estimated 17.2 million people died from communicable diseases, while 28.1 million were killed by noncommunicable illnesses, including heart disease, stroke, and cancer.[1] But by 2020, deaths from communicable diseases are expected to drop to 10.3 million, while those from noncommunicable diseases will rise to nearly 50 million worldwide—outnumbering communicable diseases five-to-one.[2] (See Table 1.) In developing countries, they will account for 7 out of every 10 deaths and nearly 60 percent of all illness and disability.[3]

Within 25 years, tobacco-induced illnesses are expected to overtake all communicable diseases as the leading threat to human health worldwide.[4] Developing countries are especially at risk because more people are smoking and, on average, each person is smoking more manufactured cigarettes—a potent and concentrated form of tobacco.[5]

Although lung cancer is the first illness to come to mind when smoking is mentioned, tobacco use actually kills more people from heart disease, stroke, and other illnesses than from lung cancer. In the United States, for example, 434,000 people died in 1993 from tobacco-related illnesses: one fourth were killed by lung cancer, but nearly half died from heart disease and stroke.[6]

Noncommunicable diseases have long been recognized as killers in industrial countries—accounting for 9 out of 10 deaths in 1990.[7] Yet 60 percent of the world's 6 million cancer deaths that year occurred in developing countries.[8] By 2010, the number of new cancer cases will double in these nations even if current rates of incidence remain the same, solely because of projected increases in the number of people over age 60.[9]

But lifestyle changes and urbanization make it likely that the incidence of cancer will increase even more than that. The same is true for other lifestyle-related noncommunicable diseases: aging populations will increase the total numbers, while greater exposure to risks will increase the rates.[10]

In some developing regions, the transition from predominantly communicable to noncommunicable diseases has already occurred. In Shanghai County, China, for example, heavy use of tobacco precipitated a vast shift in mortality patterns in less than 20 years. In the early 1960s, communicable disease, respiratory illness, and accidents were the leading causes of death.[11] By the end of the 1970s, cancer, stroke, and heart disease had taken over as the

TABLE 1: CURRENT AND PROJECTED CAUSES OF DEATHS WORLDWIDE, 1990 AND 2020

CAUSE OF DEATH	1990	2020[1]
	(million)	
Communicable Diseases[2]	17.24	10.31
Noncommunicable Diseases	28.14	49.65
Cancers	6.02	12.28
Heart disease	6.26	11.11
Stroke	4.38	7.70
Lower respiratory diseases	2.94	6.37
Diabetes	0.57	0.75
Other[3]	7.97	11.44
Injuries	5.08	8.38

[1]Baseline projections. [2]Includes maternal, perinatal, and nutritional conditions. [3]Includes digestive diseases; neuropsychiatric disorders; sense organ diseases such as glaucoma; rheumatic, hypertensive, and inflammatory cardiovascular diseases; oral conditions; and other conditions.
SOURCE: Christopher J.L. Murray and Alan D. Lopez, "Global Patterns of Cause of Death and Burden of Disease in 1990, with Projections to 2020," in World Health Organization, Report of the Ad Hoc Committee on Health Research Relating to Future Intervention Options, *Investing in Health Research and Development* (Geneva: 1996).

top killers.[12] Indeed, by 1989, cardiovascular diseases—heart disease and stroke combined—were the leading causes of death nationwide.[13]

A 1990 World Bank study on the health transition in Brazil found that deaths from heart disease, stroke, cancer, and injuries rose from 38 percent of total mortality in 1960 to 54 percent in 1986.[14] This increase occurred at the same time that deaths from infectious disease declined 70 percent.[15] By 2010, noncommunicable illnesses are expected to account for nearly three fourths of all deaths in Brazil—far outnumbering infectious diseases.[16]

Although Brazil and China are further along in the health transition than some parts of the Third World, other nations are quickly catching up. Cardiovascular diseases are soon expected to be responsible for one out of four deaths in developing countries.[17] By 2020, they are expected to kill 13.8 million people in these countries alone—twice as many as in 1990.[18]

One key to halting the movement toward noncommunicable diseases is to treat the underlying causes along with the precursor conditions. Stroke, for example, is linked to hypertension, obesity, and high salt and alcohol intake, whereas heart disease is connected with a diet high in saturated fat, hypertension, diabetes, and smoking. Clarifying the risk factors helps policymakers recognize those that are unique to each disease, such as salt or saturated fat, and address the overlapping risks, such as hypertension. It also helps to target treatment and to understand the dangerous synergies among obesity, alcohol consumption, and smoking.[19]

Community groups and public health officials are responding to changing health risks by emphasizing prevention. In the Czech Republic, for instance, anti-smoking programs based in health clinics have had considerable success at helping people stop smoking. Patients treated for reconstructive heart surgery and for heart attacks in hospitals near Prague measured declines in smoking of 40 and 72 percent respectively.[20] This high rate of success was achieved through stop-smoking programs combined with lifestyle changes to reduce blood cholesterol and blood pressure levels.[21]

Taxes and prohibitions on advertising are also effective means of reducing the threat of tobacco, alcohol, and other harmful products. In India, for example, cigarette sales declined by 15 percent after the excise tax doubled in 1986.[22]

Given the added variables of rising incomes, urbanization, and aging populations, the stage for a dramatic shift in the picture of health has already been set in developing countries. It is now up to governments, individual citizens, and private companies to determine the outcome of this transition and the health of societies worldwide. If preventive measures are widely implemented, many of these early deaths can be avoided.

Maternal Mortality Remains High   Kira Schmidt

Despite decreases in population growth and mortality rates throughout most of the world, the number of women who die in the process of giving life remains high. New estimates from the World Health Organization (WHO) reveal that 585,000 women die each year during pregnancy and childbirth.[1]

WHO defines maternal mortality as the death of a woman while pregnant or within six weeks of termination of a pregnancy from causes related to or aggravated by the pregnancy or its management.[2] Its measure is expressed as the number of maternal deaths per 100,000 live births.[3]

Maternal mortality is one development indicator that reveals particularly alarming disparities between industrial and developing countries.[4] More than 99 percent of the 585,000 maternal deaths each year occur in developing countries.[5] Maternal mortality ratios range from 870 in Africa to 390 in Asia and 190 in Latin America, whereas the ratios for Europe and North America are 36 and 11, respectively.[6] (See Table 1.) The highest ratios are found in Sierra Leone (1,800) and Afghanistan (1,700), while the lowest occur in Switzerland, Norway, and Canada (6).[7]

In many developing countries, maternal mortality is the leading cause of death of women, claiming the lives of one fourth to one half of all women of reproductive age (15–49 years old).[8] These women face a lifetime risk of dying from childbirth and pregnancy-related causes that is often 50 to more than 100 times greater than the risk faced by women in industrial countries.[9]

These figures most likely underestimate the true scope of the problem because many maternal deaths go unreported or are misclassified as deaths due to other causes.[10] Trends in maternal mortality are therefore also difficult to determine. Yet the evidence suggests that over the past two decades, ratios have declined slightly in some Latin American and Asian countries and substantially in industrial countries.[11] In developing countries with extremely high ratios, however, particularly in sub-Saharan Africa, there has been little progress.[12] Globally, minor reductions have been achieved over the past decade, but because the number of births has also increased during this period, the total number of maternal deaths has remained largely unchanged.[13]

The death toll understates the magnitude of the problem. For every maternal death, as many as 30 women sustain oftentimes crippling and lifelong health problems related to

TABLE 1: MATERNAL MORTALITY, TOTAL AND BY REGION, 1990

REGION	MATERNAL MORTALITY RATIO (deaths per 100,000 live births)	MATERNAL DEATHS (number)	LIFETIME RISK OF MATERNAL DEATH (ratio)
Africa	870	235,000	one in 16
Asia	390	323,000	one in 65
Latin America	190	23,000	one in 130
Europe	36	3,200	one in 1,400
North America[1]	11	500	one in 3,700
Less Developed Regions	480	582,000	one in 48
More Developed Regions	27	4,000	one in 1,800
World Total	430	585,000	one in 60

[1]Does not include Mexico.
SOURCE: World Health Organization, *Revised 1990 Estimates of Maternal Mortality: A New Approach by WHO and UNICEF* (Geneva: 1996).

pregnancy. Maternal morbidity affects more than 15 million women a year, and nearly 300 million—25 percent of the adult women in developing countries—have suffered or continue to suffer from such infirmities.[14]

Three quarters of maternal deaths are attributable to five direct obstetric causes: hemorrhage, infection, toxemia, obstructed labor, and complications resulting from unsafe abortion.[15] Most of these deaths occur because many women lack access to prenatal, emergency obstetric, and postnatal care to detect and treat complications of pregnancy, or because they lack access to family planning services and safe, legal abortion. The remaining 25 percent of maternal deaths result from preexisting illnesses or conditions that are exacerbated by pregnancy, such as hypertension, hepatitis, and anemia.[16]

While these are the medical causes of maternal death, the roots of the problem run much deeper. Too often, women in developing countries are poor, malnourished, and uneducated, and they lack access to health care and family planning services.[17] These women are also particularly susceptible due to relatively high numbers of pregnancies, pregnancies at particularly early and late ages, and short intervals between births.[18] All these factors compound the risk that a woman will die during pregnancy or childbirth.

The tragedy of this problem lies in the fact that nearly all these deaths are easily preventable. The medical technology to enable women to have safe births or to avoid pregnancy and childbirth has been available for decades. But in most developing countries, access to health services that provide this technology is limited by financial constraints, and, particularly in rural areas, by long distances to service providers, shortages of trained personnel, and the often poor quality of existing maternal health services and providers.[19] The World Bank estimates that an investment of less than $2 per person in developing countries could reduce maternal mortality levels by half in just one decade.[20]

Until recently, maternal mortality has not received the attention that a problem of this magnitude merits. In the past decade, however, several international conferences and initiatives have raised the profile of this largely neglected issue. The 1994 International Conference on Population and Development, for example, called for a reduction of maternal mortality to one half of 1990 levels by the year 2000 and by a further one half by 2015.[21] The Safe Motherhood Initiative, launched in 1987 by WHO, the World Bank, the U.N. Population Fund, and several international nongovernmental organizations, continues to mobilize action around the world to improve maternal health.[22]

These initiatives focus on a number of different strategies to reduce maternal mortality and the lifelong health problems that often accompany pregnancy and childbirth in the developing world. They strive to increase the provision of family planning services, access to safe abortion services, and the availability of emergency obstetric care. Efforts are also concentrated on training birth attendants, extending access to prenatal care, and promoting improvements in women's education and socioeconomic status.[23]

Many countries have begun to formulate and implement action plans to address the problem of maternal mortality.[24] Because this is such a complex and multifaceted problem, however, real progress in making motherhood safer will require political will and comprehensive efforts to tackle the fundamental problems of poverty and discrimination faced by the world's women.

Half of Languages Becoming Extinct

Hal Kane

By one estimate, between 10,000 and 15,000 languages once existed simultaneously in the world.[1] Today, only about 6,000 survive.[2] And about half of these are no longer being learned by children, so they are likely to become extinct within the next century.[3] (See Table 1.) Linguist Michael Krauss, one of the world's leading chroniclers of existing and passing languages, says this could even mean the demise or near demise of 90 percent of languages.[4]

About half of all languages today are spoken by fewer than 5,000–6,000 people, but some 200 languages are spoken by more than a million people each.[5] So many countries have English, French, Spanish, Arabic, or Portuguese as their main or preferred language that others are preferred in only about 70 countries, according to biologist Jared Diamond.[6] At best, a few hundred languages are officially protected anywhere in the world because they are adopted by governments. The remaining 5,700 or so tongues are not secure.[7]

Here are a few of the languages being lost: Ainu, formerly spoken on Hokkaido, Sakhalin, and the Kurile Islands, may now be extinct.[8] Of North American Indian languages, Iowa and Osage have just 5 fluent speakers each, Mandan has 6, and Abenaki-Penobscot has 20.[9] When in 1989 linguists discovered an elderly woman in Turkey who spoke Ubykh, a northwest Caucasian language with the most consonants ever recorded, she became the second known living speaker.[10]

Australia is likely to lose 90 percent of the 250 Aboriginal languages spoken there.[11] They are all moribund, and most are very near extinction: just 100 are still in use, and only 7 are spoken by more than 1,000 individuals.[12] Jared Diamond believes that within our lifetimes, only 2 or 3 of these will retain their vitality.[13] On average, one last speaker of an Aboriginal language dies each year in Australia.[14]

Around 300 of the 900 languages in the western hemisphere will soon disappear forever, according to Michael Krauss.[15] Some 50 indigenous languages in Central America (17 percent) and 110 of those in South America (27 percent) may be moribund and will vanish when those now speaking them die.[16]

In North America, children are no longer learning 149 of 187 native languages.[17] About 50 of these dialects are spoken in California, which is the world's third most linguistically diverse region, after New Guinea and the Caucasus.[18] Even Navajo—by far the largest native language group in North America, with 200,000 speakers—appears to be in trouble.[19] A generation ago, 90 percent of Navajo children entering school spoke the language; today, only 10 percent do.[20] In 1994, in Oklahoma, the last living speakers of Miami, Peoria, and Quapaw passed away.[21]

Alaska had 20 native Eskimo and Indian languages.[22] Eighteen of them, however, are either moribund or have disappeared. Only Siberian Yupik, with 1,000 speakers, and Central Yupik of St. Lawrence Island, with 10,000 speakers, are still being learned by children.[23]

Western Europe has the fewest native languages—45, far fewer than Australia's 250 even though Australia has a much smaller population.[24] This is because the expansion of empires and the spread of agriculture and commerce bring people together under one or a few languages, a process that happened long ago in Europe but never occurred in parts of Australia.[25] The expansion of Indo-European farmers and herders that began around 4000 B.C. eradicated all existing West European languages except Basque.[26] The Americas had more than 1,000 languages before Europeans arrived, but now Spanish, English, and Portuguese dominate.[27]

Some 3,500 languages are found in just nine countries: Papua New Guinea has 850, Indonesia had 670, Nigeria has 410, India has 380, Cameroon has about 270, Australia has some 250, Mexico has about 240, and Zaire and Brazil each have about 210.[28] And the tiny nation of Vanuatu alone has about 105.[29] Many New Guinean languages are so distinctive that they have no proven relationship with any other in the world.[30] New Guinea, Vanuatu, the Philippines, and Australia were never unified by an empire before the Europeans came. Today, most of their languages are spoken by fewer than 1,000 people.[31]

As humanity's linguistic heritage disappears, one language is coming to dominate the world. English now has more nonnative speakers than native ones. Some estimates put the ratio at four-to-one in favor of the nonnative speakers.[32] This means that most English is spoken by one nonnative speaker to another. And in classes, English is taught mostly by nonnative speakers to nonnative speakers.[33] Linguists say "the sun now sets on the British Empire, but never on the English language."[34]

One out of every six Chinese is studying English, adding up to more than 200 million people.[35] Worldwide, one out of five people speak at least a small amount of English, and three quarters of the world's mail is written in this language.[36] Some 80 percent of the electronically stored information is in English.[37] And more than two thirds of the world's scientists read in this dominant tongue.[38]

Countering this trend toward homogenization is the fact that one dead language has been revived fully, and one other now has a handful of speakers. The first native speaker of Hebrew in 1,500 years was recorded in the late nineteenth century, after being raised to speak this as his first language.[39] Today, of course, Hebrew is the national language of Israel. And Cornish, after becoming extinct in the late eighteenth century, was reconstituted from the texts of fourteenth-century plays; today there are several young native speakers in Cornwall.[40]

Different languages carry the diversity of human experience, and they facilitate the kind of thinking that responds to the experiences and needs of many cultures. To lose those languages is to lose a part of human experience and, possibly, to constrain some of the range of our thinking.

TABLE 1: PARTIAL SURVEY OF MORIBUND LANGUAGES, BY REGION[1]

LOCATION	LANGUAGES	NUMBER OR SHARE ENDANGERED
Alaska	20	18 languages
Alaska and the Soviet North	50	45 languages
Russia	65	45 languages
United States and Canada	187	149 languages
Australia	250	90 percent
Meso-America (incl. Mexico)	300	50 languages
South America	400	110 languages
All the Americas	900	300 languages
All former Soviet Union	n.a.	50 percent
Europe	Many European languages went extinct long ago; a few are currently going extinct	
World	6,000	3,000 languages approximately

[1]Language statistics are rare. Many languages have not been catalogued at all or well; even for the well known ones, the number of speakers is hard to count and the number of young speakers is even harder to know. Government statistics are often not reliable because the government may favor one language over others.
SOURCE: Michael Krauss, "The World's Languages in Crisis," *Language* 68, no. 1 (1992).

NOTES

WORLD GRAIN HARVEST SETS RECORD (pages 26–27)

1. U.S. Department of Agriculture (USDA), Foreign Agricultural Service (FAS), *Grain: World Markets and Trade* (Washington, DC: January 1997).
2. USDA, *Production, Supply, and Distribution* (PS&D), electronic database, Washington, DC, updated November 1996; USDA, "World Grain Database," unpublished printout, Washington, DC, 1991; USDA, op. cit note 1.
3. USDA, op. cit. note 1; USDA, *PS&D*, op. cit. note 2.
4. USDA, op. cite. note 1; USDA, *PS&D*, op. cit. note 2; "Future Prices," *Wall Street Journal*, various editions.
5. USDA, op. cit. note 1; USDA, *PS&D*, op. cit. note 2; Gary Gardner, *Shrinking Fields: Cropland Loss in a World of Eight Billion*, Worldwatch Paper 131 (Washington, DC: Worldwatch Institute, July 1996).
6. Gardner, op. cit. note 5.
7. USDA, FAS, *World Agricultural Production* (Washington, DC: October 1995).
8. USDA, FAS, *Oilseeds: World Markets and Trade* (Washington, DC: December 1996).
9. David Seckler, *The New Era of Water Resources Management: From "Dry" to "Wet" Water Savings*, Issues in Agriculture 8 (Washington, DC: Consultative Group on International Agricultural Research, April 1996).
10. K.G. Soh and K.F. Isherwood, "Short Term Prospects for World Agriculture and Fertilizer Use," presentation at IFA Enlarged Council Meeting, International Fertilizer Industry Association, Marrakech, Morocco, 19–22 November 1996.
11. K.F. Isherwood and K.G. Soh, "Short Term Prospects for World Agriculture and Fertilizer Use," presentation at 21st IFA Enlarged Council Meeting, International Fertilizer Industry Association, Cape Town, South Africa, 15–17 November 1995.
12. USDA, op. cit. note 1; USDA, *PS&D*, op. cit. note 2; U.S. Bureau of Census, *International Data Base*, electronic database, Suitland, MD, updated 15 May 1996.
13. USDA op. cit. note 1; USDA, *PS&D*, op. cit. note 2; Bureau of Census, op. cit. note 12.
14. USDA op. cit. note 1; Bureau of Census, op. cit. note 12.
15. USDA, op. cit. note 1; USDA, *PS&D*, op. cit. note 2.
16. USDA, op. cit. note 1; USDA, *PS&D*, op. cit. note 2.
17. USDA, op. cit. note 1; USDA, *PS&D*, op. cit. note 2; USDA, *Agricultural Statistics 1995–96* (Washington, DC: U.S. Government Printing Office, 1995–96).
18. USDA, op. cit. note 1.
19. "EU Gears to Expand Grain Production Next Year," in USDA, FAS, *Grain: World Markets and Trade* (Washington, DC: November 1995); Alan Rifkin, FAS, USDA, Washington, DC, discussion with Gary Gardner, 6 February 1997. Set-asides in Europe respond to internal European Union agricultural policy and may be unrelated to global trends.

SOYBEAN HARVEST RECOVERS TO NEAR-RECORD (pages 28–29)

1. U.S. Department of Agriculture (USDA), Foreign Agricultural Service (FAS), *Oilseeds: World Markets and Trade* (Washington, DC: December 1996).
2. Ibid.; USDA, *Production, Supply, and Distribution*, electronic database, Washington, DC, updated November 1996.
3. USDA, op. cit. note 1; U.S. Bureau of Census, *International Data Base*, electronic database, Suitland, MD, updated 15 May 1996.
4. USDA, op. cit. note 1; USDA, op. cit. note 2; Bureau of Census, op. cit. note 3.

5. USDA, op. cit. note 1.
6. Ibid.
7. Ibid.; USDA, op. cit. note 2.
8. USDA, op. cit. note 1.
9. Ibid.
10. McVean Trading and Investments, Memphis, TN, discussion with author, 13 September 1996.
11. Ibid.
12. Ibid.
13. Ibid.
14. USDA, Economic Research Service, *Agricultural Outlook* (Washington, DC: September 1996).
15. Ibid.; USDA, op. cit. note 1.
16. USDA, op. cit. note 1.
17. Ibid.
18. McVean, op. cit. note 10.

MEAT PRODUCTION GROWTH SLOWS (pages 30–31)

1. U.S. Department of Agriculture (USDA), Foreign Agricultural Service (FAS), *Livestock and Poultry: World Markets and Trade* (Washington, DC: October 1996).
2. Ibid.
3. "Future Prices," *Wall Street Journal*, various editions.
4. USDA, op. cit. note 1.
5. Ibid.
6. Ibid.
7. Ibid.
8. Compiled by Worldwatch Institute from U.N. Food and Agriculture Organization (FAO), *1948–1985 World Crop and Livestock Statistics* (Rome: 1987); FAO, *FAO Production Yearbooks 1988–1991* (Rome: 1991–1993); USDA, op. cit. note 1.
9. USDA, op. cit. note 1.
10. Ibid.
11. Ibid.
12. Ibid.
13. Ibid.
14. Ibid.
15. USDA, FAS, *World Agricultural Production* (Washington, DC: October 1996).
16. Ibid.
17. Ibid.
18. USDA, op. cit. note 1.
19. USDA, op. cit. note 1; "Future Prices," op. cit. note 3.
20. USDA, op. cit. note 1.
21. USDA, op. cit. note 15.
22. USDA, op. cit. note 1.
23. Ibid.
24. Ibid.
25. USDA, op. cit. note 15.

GLOBAL FISH CATCH REMAINS STEADY (pages 32–33)

1. Data for 1990–95 from Maurizio Perotti, fishery statistician, Fishery Information, Data and Statistics Unit (FIDI), Fisheries Department, U.N. Food and Agriculture Organization (FAO), Rome, letter to author, 8 November 1996; 1950–90 total catch from FIDI, *Yearbook of Fishery Statistics: Catches and Landings* (Rome: FAO, 1967–91).
2. FIDI, op. cit. note 1; 1990–95 data from Perotti, op. cit. note 1; population data from U.S. Bureau of Census, *International Data Base*, electronic database, Suitland, MD, 15 May 1996.
3. Perotti, op. cit. note 1; 1984–94 culture data from FAO, *Aquaculture Production Statistics, 1985–1994*, FAO Fisheries Circular No. 815, Rev. 8 (Rome: 1996); before 1984, culture estimates are based on a 1975 aquaculture production estimate from National Research Council, *Aquaculture in the United States: Constraints and Opportunities* (Washington, DC: National Academy of Sciences, 1978) and country estimates in Conner Bailey and Mike Skladany, "Aquacultural Development in Tropical Asia," *Natural Resources Forum*, February 1991; growth rates from FAO, *Marine Fisheries and the Law of the Sea: A Decade of Change*, FAO Fisheries Circular No. 853 (Rome: 1993).
4. Perotti, op. cit. note 1.
5. Bob Holmes, "Blue Revolutionaries," *New Scientist*, 7 December 1996.
6. FIDI, op. cit. note 1.
7. Xinhua News Agency, "More Fish to Be Reared as Part of Food Solution," *China Daily*, 18 October 1996.
8. Ibid.
9. Biksham Gujja and Andrea Finger-Stich, "What Price Prawn? Shrimp Aquaculture's Impact in Asia," *Environment*, September 1996.
10. Meryl Williams, *The Transition in the Contribution of Living Aquatic Resources to Food Security*, Food, Agriculture, and the Environment Discussion Paper 13 (Washington, DC: International Food Policy Research Institute, April 1996).

11. John Tibetts, "What's the Catch? Seafood's New Frontier," *Coastal Heritage*, Fall 1996.
12. Gujja and Finger-Stich, op. cit. note 9.
13. Mathew Gianni, Greenpeace, "Non-Governmental Organization (NGO) Statement on Unsustainable Aquaculture," Association for Progressive Communications conference <env.marine> (maintained and archived by the Institute for Global Communications, San Francisco), posted 6 May 1996.
14. Figure of 80 percent based on Perotti, op. cit. note 1.
15. "News Round-Up: Dwindling," *Samudra* (Madras, India: International Collective in Support of Fishworkers), July 1996.
16. National Marine Fisheries Service, National Oceanic and Atmospheric Administration, U.S. Department of Commerce, *Our Living Oceans: Report on the Status of U.S. Living Marine Resources, 1995* (Silver Spring, MD: February 1996).
17. Janet Raloff, "Fishing for Answers: Deep Trawls Leave Destruction in Their Wake—But for How Long?" *Science News*, 26 October 1996.
18. Ibid.
19. Caroline Southey, "Brussels Retreats on Cuts to Fishing Fleets," *Financial Times*, 11 October 1996.
20. Ibid.; Caroline Southey, "Move for Large Cuts in Fish Catches Undermined," *Financial Times*, 22 November 1996.
21. Williams, op. cit. note 10.

GRAIN STOCKS UP SLIGHTLY
(pages 34–35)

1. U.S. Department of Agriculture (USDA), Foreign Agricultural Service (FAS), *Grain: World Markets and Trade* (Washington, DC: January 1997).
2. Ibid.; USDA, *Production, Supply, and Distribution*, electronic database, Washington, DC, updated November 1996.
3. USDA, op. cit. note 1.
4. Ibid.
5. Ibid.
6. Ibid.; USDA, op. cit. note 2.
7. USDA, op. cit. note 1.
8. Ibid.
9. Ibid.; USDA, op. cit. note 2.
10. USDA, op. cit note 1.
11. Hunter Colby, Fredrick W. Crook, and Shwu-

Eng H. Webb, *Agricultural Statistics of the People's Republic of China, 1949–1950* (Washington, DC: USDA, Economic Research Service, 1992).
12. James Hansen et al., Goddard Institute for Space Studies Surface Air Temperature Analyses, "Global Land-Ocean Temperature Index," <http://www.giss.nasa.gov./Data/GIS-TEMP>, viewed 14 January 1997.
13. Ibid.
14. USDA, op. cit. note 1; USDA, "Weekly Weather and Crop Bulletin" <gopher://mann77.mannlib.cornell.edu:70/11/reports/nassr/field/weather>, Washington, DC, various dates.
15. U.S. Bureau of the Census, *International Data Base*, electronic database, Suitland, MD, updated 15 May 1996.

FERTILIZER USE RISING AGAIN
(pages 38–39)

1. K.G. Soh and K.F. Isherwood, "Short Term Prospects for World Agriculture and Fertilizer Use," presentation at IFA Enlarged Council Meeting, International Fertilizer Industry Association, Marrakech, Morocco, 19–22 November 1996.
2. K.G. Soh and K.F. Isherwood, "The Agricultural Situation and Fertilizer Demand," presentation at 64th IFA Annual Conference, International Fertilizer Industry Association, Agro-Economics Committee, Berlin, 20–23 May 1996.
3. Soh and Isherwood, op. cit. note 1.
4. Ibid.
5. Soh and Isherwood, op. cit. note 2.
6. Soh and Isherwood, op. cit. note 1.
7. Ibid.
8. Ibid.
9. Ibid.
10. U.N. Food and Agriculture Organization (FAO), *Fertilizer Yearbook* (Rome: various years); Soh and Isherwood, op. cit. note 1.
11. FAO, op. cit. note 10; Soh and Isherwood, op. cit. note 1.
12. FAO, op. cit. note 10; Soh and Isherwood, op. cit. note 1.
13. FAO, op. cit. note 10.
14. K.F. Isherwood and K.G. Soh, "Short Term Prospects For World Agriculture and Fertilizer Use," presentation at 21st IFA Enlarged Council Meeting, International Fertilizer Industry Association, Cape Town, 15–17 November

1995.

15. FAO, op. cit. note 10; Soh and Isherwood, op. cit. note 1.

16. Soh and Isherwood, op. cit. 1.

17. FAO, *Production Yearbook 1990* (Rome: 1991); Bill Quinby, Economic Research Service, U.S. Department of Agriculture (USDA), Washington, DC, letter to Gary Gardner, 24 January 1996.

18. David Seckler, *The New Era of Water Resources Management: From "Dry" to "Wet" Water Savings,* Issues in Agriculture 8 (Washington, DC: Consultative Group on International Agricultural Research, April 1996).

19. Lester R. Brown and Hal Kane, *Full House: Reassessing the Earth's Carrying Capacity* (New York: W.W. Norton & Company, 1994).

20. Isherwood and Soh, op. cit 14.

21. Ibid.

22. Ibid.

23. USDA, Foreign Agricultural Service, *Grain: World Markets and Trade* (Washington, DC: December 1996).

GRAIN AREA JUMPS SHARPLY
(pages 40–41)

1. U.S. Department of Agriculture (USDA), Foreign Agricultural Service (FAS), *Grain: World Markets and Trade* (Washington, DC: January 1997).

2. Surge from ibid. and from USDA, *Production, Supply, and Distribution,* electronic database, Washington, DC, updated February 1996.

3. USDA, op. cit. note 2; population figures from U.S. Bureau of the Census, *International Data Base,* electronic database, Suitland, MD, updated 15 May 1996.

4. Tim Dyson, *Population and Food: Global Trends and Future Prospects* (London: Routledge, 1996).

5. USDA, op. cit. note 2.

6. Ibid.

7. Ibid. Until the twentieth century, area expansion was the chief tool for meeting the annual increase in food demand. Therefore, grain area can be assumed to have increased in step with global population growth since the dawn of agriculture 10,000 years ago.

8. See Gary Gardner, *Shrinking Fields: Cropland Loss in a World of Eight Billion,* Worldwatch Paper 131 (Washington, DC: Worldwatch Institute, July 1996); for an alternative view, see U.N. Food and Agriculture Organization, *World Agriculture: Towards 2010* (New York: John Wiley & Sons, 1995).

9. Tim Warman, "1996 Farm Bill: A Triumph for Conservation," *American Farmland,* spring 1996; idled land from USDA, Economic Research Service, "AREI Updates: 1995 Cropland Use," Number 12, Washington, DC, 1995.

10. "EU Gears to Expand Grain Production Next Year," in USDA, FAS, *Grain: World Markets and Trade* (Washington, DC: November 1995); Alan Rifkin, FAS, USDA, Washington, DC, discussion with author, 6 February 1997. Set-asides in Europe respond to internal European Union agricultural policy and may be unrelated to global trends.

11. As grain area contracted in most years since 1981, soybean acreage increased by 22 percent, and area increases of more than 15 percent were registered for nine major fruits and vegetables; Gardner, op. cit. note 8.

12. Gardner, op. cit. note 8.

13. "Two Decades of Growth in the Production of World Grains and Oilseeds," in USDA, FAS, *World Agricultural Production* (Washington, DC: October 1995).

14. Gardner, op. cit. note 8.

15. Sara J. Scherr and Satya Yadav, *Land Degradation in the Developing World: Implications for Food, Agriculture, and the Environment to 2020,* Food, Agriculture, and the Environment Discussion Paper 14 (Washington, DC: International Food Policy Research Institute, May 1996); L.R. Oldeman, *World Map of the Status of Human-Induced Soil Degradation: An Explanatory Note,* 2nd ed. (Wageningen, Netherlands, and Nairobi: International Soil Reference and Information Centre and United Nations Environment Programme, 1991).

16. USDA, op. cit. note 2.

17. Ibid.

18. International Monetary Fund, *World Economic Outlook* (Washington, DC: May 1996).

19. China from George P. Brown, "Arable Land Loss in Rural China," *Asian Survey,* October 1995.

20. Consultative Group on International Agricultural Research, "Poor Farmers Could Destroy Half of Remaining Tropical Forest," press release (Washington, DC: 4 August 1996).

IRRIGATED AREA UP SLIGHTLY
(pages 42–43)

1. U.N. Food and Agriculture Organization (FAO), "FAOSTAT DATA," < http://www.fao.org >, Rome, viewed 8 October 1996, with adjustments to U.S. data from Bill Quinby, Economic Research Service (ERS), U.S. Department of Agriculture (USDA), Washington, DC, letter to author, 24 January 1996.
2. Worldwatch calculation based on irrigated area from FAO, op. cit. note 1, and on population data from U.S. Bureau of the Census, as made available in *World Population, Midyear 1950–2050*, electronic database, ERS, USDA, Washington, DC, updated July 1995.
3. Worldwatch calculation based on irrigated area from FAO, op. cit. note 1, and on population data from Bureau of the Census, op. cit. note 2.
4. FAO, *State of Food and Agriculture 1993* (Rome: 1993); FAO, op. cit. note 1.
5. Worldwatch calculation based on irrigated area from FAO, op. cit. note 1; and on population data from Bureau of the Census, op. cit. note 2.
6. FAO, op cit. note 1.
7. Worldwatch calculation based on irrigated area from FAO, op. cit. note 1; and on population data from Bureau of the Census, op. cit. note 2.
8. FAO, op. cit. note 1.
9. Sandra Postel, "Forging a Sustainable Water Strategy," in Lester R. Brown et al., *State of the World 1996* (New York: W.W. Norton & Company, 1996).
10. FAO, "Food Production: The Critical Role of Water," in *Technical Background Documents 6–11: Vol. 2* (Rome: 1996).
11. Pierre Crosson, "Future Supplies of Land and Water for World Agriculture," revision of a paper presented at a conference of the International Food Policy Research Institute in February 1994, Resources for the Future, Washington, DC, August 1994.
12. Sandra Postel, *Dividing the Waters: Food Security, Ecosystem Health, and the New Politics of Scarcity,* Worldwatch Paper 132 (Washington, DC: Worldwatch Institute, September 1996).
13. FAO, *State of Food and Agriculture 1993* (Rome: 1993).
14. Robert Goodland, "Environmental Sustainability Needs Renewable Energy: The Extent to Which Big Hydro is Part of the Transition," presentation at International Crane Foundation Workshop, Washington, DC, 28 November–2 December 1995.
15. Postel, op. cit. note 12.
16. Gary Gardner, *Shrinking Fields: Cropland Loss in a World of Eight Billion,* Worldwatch Paper 131 (Washington, DC: Worldwatch Institute, July 1996).
17. Department of Conservation, State of California, *Farmland Conversion Report,* 1990–92 (Sacramento: 1994).
18. Worldwatch calculation based on data in ibid.
19. Gardner, op. cit. note 16.
20. Ibid.
21. Paul Raskin, Evan Hansen, and Robert Margolis, *Water and Sustainability: A Global Outlook,* Polestar Series Report No. 4 (Stockholm: Stockholm Environment Institute, 1995).

FOSSIL FUEL USE SURGES TO NEW HIGH (pages 46–47)

1. Data for 1950–95 from United Nations, *World Energy Supplies* (New York: 1976), from United Nations, *Yearbook of World Energy Statistics* (New York: 1983), from United Nations, *Energy Statistics Yearbook* (New York: various years), from U.S. Department of Energy (DOE), Energy Information Administration (EIA), *Annual Energy Review 1995* (Washington, DC: July 1996), and from DOE, EIA, *International Energy Annual 1995* (Washington, DC: December 1996); 1996 figure is Worldwatch estimate based on British Petroleum (BP), *BP Statistical Review of World Energy* (London: Group Media & Publications, 1996), on DOE, EIA, *Monthly Energy Review January 1997* (Washington, DC: 1997), on P.T. Bangsberg, "Bleak Future Envisioned for China's Coal Sector," *Journal of Commerce,* 30 December 1996, on "Worldwide Look at Reserves and Production," *Oil & Gas Journal,* 25 December 1995, on PlanEcon Inc., *PlanEcon Energy Outlook, Eastern Europe and Former Soviet Union* (Washington, DC: 1997), and on V. Luque Cabal, European Commission, Energy Directorate, letter to author, 29 January 1997.
2. Worldwatch estimate based on United Nations, *Energy Statistics Yearbook* (New York: 1996).
3. PlanEcon, op. cit. note 1.
4. Worldwatch estimate based on BP, op. cit. note 1, on DOE, *Monthly Energy Review,* op. cit. note 1, and on PlanEcon, op. cit. note 1.
5. DOE, *Monthly Energy Review,* op. cit. note 1.
6. Worldwatch estimate based on BP, op. cit. note

1, on DOE, *Monthly Energy Review*, op. cit. note 1, and on PlanEcon, op. cit. note 1.

7. "Worldwide Look at Reserves and Production," op. cit. note 1.

8. BP, op. cit. note 1; "Worldwide Look at Reserves and Production," op. cit. note 1.

9. "Worldwide Look at Reserves and Production," op. cit. note 1.

10. Ibid.

11. James J. MacKenzie, *Oil as a Finite Resource: When is Global Production Likely to Peak?* (Washington, DC: World Resources Institute, March 1996).

12. Worldwatch estimate based on DOE, *Monthly Energy Review*, op. cit. note 1, on Bangsberg, op. cit. note 1, on BP, op. cit. note 1, on Cabal, op. cit. note 1, and on PlanEcon, op. cit. note 1.

13. Cabal, op. cit. note 1; BP, op. cit. note 1.

14. BP, op. cit. note 1.

15. Ibid.

16. DOE, *Monthly Energy Review,* op. cit. note 1.

17. Bangsberg, op. cit. note 1.

18. Ibid.

19. Worldwatch estimate based on DOE, *Monthly Energy Review*, op. cit. note 1, on PlanEcon, op. cit. note 1, on BP, op. cit. note 1, and on Cabal, op. cit. note 1.

20. BP, op. cit. note 1.

21. PlanEcon, op. cit. note 1.

22. Chang Wimin, "China to Import Gas from Central Asia, Russia," *China Daily*, 5 February 1996.

23. Ibid.

24. Eurogas, "10% Growth of Natural Gas Consumption in 1996 in Western Europe," press release (Brussels: 10 February 1997).

25. Robert Corzine and Neil Buckley, "On the Front Burner: After Decades of Monopolistic Control, Europe's Gas Market is Poised for Liberalisation," *Financial Times*, 20 November 1996.

NUCLEAR POWER INCHES UP
(pages 48–49)

1. Installed nuclear capacity is defined as reactors connected to the grid as of 31 December 1996, and is based on Worldwatch Institute database complied from statistics from the International Atomic Energy Agency (IAEA) and press reports, primarily from *European Energy Report*, *Energy Economist*, *Nuclear News*, *New York Times*, *Journal of Commerce*, *Financial Times*, and World Wide Web sites.

2. Worldwatch Institute database, op. cit. note 1.

3. Ibid.

4. Ibid.

5. Ibid.

6. Ibid.

7. Ibid.

8. "TVA's Watts Bar Unit-1 Starting Up; Maine Yankee Recommencted to Grid," *McGraw-Hill Companies' Electric Power Daily*, 22 January 1996.

9. "Ukraine Closes One of Two Working Reactors At Chernobyl," *The Energy Daily*, on Newspage < http://pnp.individual.com/PNP/pnp. frames.-html >, posted 5 December 1996.

10. "Russian N-Plant Cash Crisis Deepens," *Nucnet* < http://www.aey.ch/nucnet >, posted 23 September 1996.

11. "Russian Region Votes 'No' to Nuclear Plant," *Reuters*, on Energy Central < http://www.energycentral.com >, posted 9 December 1996.

12. Worldwatch Institute database, op. cit. note 1.

13. "Closing Dutch Reactor," *Sustainable Energy News*, November 1996.

14. "Flotation of British Nuclear Firm Nets Disappointing $2.24-Billion," *Electric Utility Week*, 22 July 1996; "Nuclear Sell-off Plan Slated," *European Energy Report*, 24 November 1995.

15. "U.K. Government Prepares Sale of 8,500 MW of Nuclear Assets," *Electric Utility Week*, 3 June 1996.

16. Worldwatch Institute database, op. cit. note 1.

17. Ibid.

18. "Plebiscite Rejecting Nuclear Power Project May Force Country to Alter Energy Policies," *International Environment Reporter*, 21 August 1996.

19. "Site of Japan Nuclear Reactor Rejects Subsidies," *Reuters*, on Energy Central < http://www.energycentral.com >, posted 20 November 1996.

20. "ABB Unit Awarded Contracts Worth About $200 Million for Korean Nuclear Power Plants," *Business Wire*, 25 November 1996; Anna Gyorgy, Green Korea, Seoul, e-mail message to author, 6 January 1997.

21. "Fact Sheet on Nuclear Power in South Korea," *Green Korea Reports*, January 1997; Gyorgy, op. cit. note 20.

22. "Electrifying Asia: Demand Faces Resource Constraints," *Energy Economist Briefings*, July 1996.

23. James Yarley, "Daya Bay and Beyond," *Nuclear Engineering International*, December 1996.
24. Lis Tacey, "Taiwan Finally Clears Nuclear Power Plant," *Nature*, 24 October 1996; Howard Benowitz, Scott Denman, and Nicholas Lenssen, *International Nuclear Power: Mythbusters #10* (Washington, DC: Safe Energy Communications Council, Spring 1996).
25 Tacey, op. cit. note 24.
26 Ercan Ersoy, "Turkey Tenders for First Nuclear Plant, Sees More," *Reuters*, on Energy Central < http://www.energycentral.com >, posted 18 December 1996.

GEOTHERMAL POWER RISES
(pages 50–51)

1. Figures for 1950–94 from Mary H. Dickson and Mario Fanelli, "Geothermal Energy Worldwide," in *The World Directory of Renewable Energy Suppliers and Services 1995* (London: James & James Science Publishers Ltd., 1995); 1995–96 figures from Mary Dickson, International Institute for Geothermal Research, Pisa, Italy, letter to author, 3 February 1997.
2. Gerald W. Huttrer, "The Status of World Geothermal Power Production 1990–1994," presentation at World Geothermal Conference, Florence, May 1995.
3. Priscilla Ross, "Hot Rocks," *Energy Economist*, September 1995; figure of 1 percent is based on data on total world electricity generation in U.S. Department of Energy, Energy Information Administration, *International Energy Outlook 1996* (Washington, DC: 1996).
4. David Tenenbaum, "Tapping the Fire," *Technology Review*, January 1995.
5. Ibid.
6. Ronald DiPippo, "Geothermal Power Plants in the United States: A Survey and Update for 1990–1994," *GRC Bulletin*, May 1995; "Geothermal: Oregon Project Has Uncertain Future; Developer Tries to Move to California," *International Solar Energy Intelligence Report*, 1 November 1996.
7. Huttrer, op. cit. note 2; "Philippines Geothermal Power Plant," *South*, November 1996.
8. Huttrer, op. cit. note 2.
9. Ibid.
10. Ibid.
11. Tenenbaum, op. cit. note 4.
12. Huttrer, op. cit. note 2.
13. Ibid.
14. Dickson and Fanelli, op. cit. note 1.
15. Ibid.
16. Ibid.
17. Ibid.
18. Civis G. Palmerini, "Geothermal Energy," in Thomas B. Johansson et al., eds., *Renewable Energy: Sources for Fuels and Electricity* (Washington, DC: Island Press, 1993); Huttrer, op. cit. note 2.
19. Frank Pfaff, "German Town Uses Geothermal Power," *Deutsche Presse-Agentur*, 20 December 1996; Huttrer, op. cit. note 2; Dickson and Fanelli, op. cit. note 1.
20. Tenenbaum, op. cit. note 4.
21. Ibid.
22. Ibid.
23. Sheila Polson, "Geothermal Pumps Tap Energy of the Earth Year-Round," *Christian Science Monitor*, 28 May 1996; Dickson and Fanelli, op. cit. note 1.
24. Dickson and Fanelli, op. cit. note 1.
25. Gerald Huttrer, Geothermal Management Company, Inc., Frisco, CO, discussion with author, 15 January 1997; "North Island, N.Z. Geothermal Field Near Lake Taupo Eyed By Three Firms," *The Solar Letter*, 13 September 1996.
26. "Philippines Geothermal Power Plant," op. cit. note 7; Tenenbaum, op. cit. note 4.

WIND POWER GROWTH CONTINUES (pages 52–53)

1. Estimates based on figures supplied by Birger Madsen, BTM Consult, "International Wind Energy Development" (Ringkobing, Denmark: 17 January 1997), by Paul Gipe and Associates, Tehachapi, CA, discussion with author, 19 February 1996, and by Knud Rehfeldt, Deutsches Windenergie-Institut, letter to author, 13 January 1997.
2. Madsen, op. cit. note 1; the total installed capacity for the end of 1996 reflects both the 1,200 megawatts of turbines added during the year and any turbines abandoned or dismantled.
3. Figure of 1 percent is based on data on total world electricity generation in U.S. Department of Energy, Energy Information Administration, *International Energy Outlook 1996* (Washington,

DC: 1996).

4. Rehfeldt, op. cit. note 1.

5. Sara Knight, "Dip in Growth Catches Firms Unawares," *Windpower Monthly*, November 1996.

6. Madsen, op. cit. note 1.

7. Neelam Matthews, "Survival of the Toughest," *Windpower Monthly*, January 1997.

8. Madsen, op. cit. note 1.

9. Hilary Barnes, "Windmill Turnover Gains Momentum in Danish Crusade," *Financial Times*, 4 April 1996.

10. "Denmark: Domestic Market Revives with 1500 MW Target," *Wind Directions*, April 1996.

11. Madsen, op. cit. note 1.

12. "Wind Energy in Spain," *Wind Directions*, January 1996.

13. Madsen, op. cit. note 1.

14. "Liberalisation Uncertainty," *Windpower Monthly*, February 1997.

15. Ibid.

16. Lyn Harrison, "China Places Major Wind Plant Orders," *Windpower Monthly*, January 1997.

17. Paul Gipe, "California Wind Industry Declining Precipitously Says New Data" (Tehachapi, CA: Paul Gipe and Associates, 16 January 1997).

18. Enron Corp, "Enron Forms Enron Renewable Energy Corp: Acquires Zond Corporation, Leading Developer of Wind Energy Power," press release (Houston, TX: 6 January 1997).

SOLAR CELL SHIPMENTS KEEP RISING (pages 54–55)

1. Paul Maycock, "1996 World Cell/Module Shipments," *PV News*, February 1997.

2. Ibid.

3. Ibid.

4. Ibid.; Paul Maycock, Photovoltaic Energy Systems, Inc., Warrenton, VA, discussion with author, January 1997.

5. Paul Maycock, "Sliced Single Crystal and Polycrystal Silicon Exceed 80 Percent Market Share. Amorphous Silicon Slips to 13%," *PV News*, February 1997.

6. Maycock, op. cit. note 1.

7. James Rannels, Office of Photovoltaic and Wind Technologies, U.S. Department of Energy, Washington, DC, discussion with author, 28 January 1996.

8. Sheri Prasso, "Technology Advances, Foreign Demand Shine New Light on Solar Industry,"

Associated Press, 16 February 1996; "Siemens Solar Completed Vancouver Crystal Growing Facility Expansion," press release (Cupertino, CA: 14 February 1996).

9. Paul Maycock, "U.S. PV Cell/Module Shipments Increase 14.7 Percent," *Photovoltaic News*, February 1997.

10. Clay Aldrich, Solar Energy Industries Association, Washington, DC, based on U.S. Department of Commerce statistics, letter to author, 31 January 1997.

11. "DOE Seeks More Active and Passive Solar, Efficient Equipment for Buildings," *The Solar Letter*, 6 December 1996; U.S. Department of Energy, "The President's Million Roofs Solar Power Initiative" (draft) (Washington, DC: 22 January 1997).

12. Maycock, op. cit. note 1.

13. "Home-use Solar Energy Systems Spreading in Japan," *Kyodo News Service*, 27 April 1996; Jochen Goebel, "Solar Energy is Gaining Momentum in Japan," *Deutsche Presse-Agentur*, 20 June 1996.

14. "Home-use Solar Energy Systems in Japan," op. cit. note 13.

15. Maycock, op. cit. note 1.

16. Paul Maycock, "European Module Shipment Flat For the First Time Ever!" *PV News*, February 1996; Paul Maycock, "European Cell/Module Shipments Decrease," *PV News*, February 1997.

17. Bob Johnson, Strategies Unlimited, Mountain View, CA, discussion with author, 31 January 1997.

18. Tom Mead, "Solar Power Takes Flight," *Financial Times*, 2 October 1996.

19. Neville Williams, Solar Electric Light Fund, Washington, DC, discussion with author, 29 January 1997.

20. Ibid.

21. Anil Cabraal, Mac Cosgrove-Davies, and Loretta Schaeffer, *Best Practices for Photovoltaic Household Electrification Programs*, World Bank Technical Paper Number 324 (Washington, DC: World Bank, 1996).

22. Michael Northrop, Peter Riggs, and Frances Raymond, "Selling Solar: Financing Household Solar Energy in the Developing World," The Pocantico Conference Center of the Rockefeller Brothers Fund, June 1996, with edited version by Michael Northrup, "Selling Solar: Financing Household Solar Energy in the Developing World," *Solar Today*, January/February 1997.

23. Rural population in India without electricity

from Kui-Nang Mak and Walter Shearer, "Sustainable Energy Development in Rural Asia," *Natural Resources Forum* 20, no. 4 (1996); "Syndicate Bank to Finance Solar Power Systems," *Deccan Chronicle*, 21 February 1996.

24. International Finance Corporation and World Bank, "The Photovoltaic Market Transformation Initiative," Background Paper (Washington, DC: October 1996); Ted Kennedy, International Finance Corporation, Washington, DC, discussion with author, 17 January 1997.

CARBON EMISSIONS SET NEW RECORD (pages 58–59)

1. The 1991 record was largely due to the Kuwaiti oil field fires, which injected 130 million tons of carbon into the atmosphere. Figures for 1950–92 (including gas flaring, but not cement production numbers) from T.A. Boden, G. Marland, and R.J. Andres, *Estimates of Global, Regional and National Annual CO₂ Emissions From Fossil Fuel Burning, Hydraulic Cement Production, and Gas Flaring: 1950–92* (Oak Ridge, TN: Carbon Dioxide Information Analysis Center, Oak Ridge National Laboratory, December 1995); 1993–95 figures are Worldwatch estimates based on ibid. and on British Petroleum (BP), *BP Statistical Review of World Energy* (London: Group Media & Publications, 1996); 1996 figure is a preliminary Worldwatch estimate based on ibid., on Boden, Marland, and Andres, op. cit. this note, on U.S. Department of Energy (DOE), Energy Information Administration (EIA), *Monthly Energy Review December 1996* (Washington, DC: 1996), on P.T. Bangsberg, "Bleak Future Envisioned for China's Coal Sector," *Journal of Commerce*, 30 December 1996, on "Worldwide Look at Reserves and Production," *Oil and Gas Journal*, 30 December 1996, on *PlanEcon Energy Outlook, Eastern Europe and Former Soviet Union* (Washington, DC: PlanEcon Inc., 1997), and on V. Luque Cabal, European Commission, Energy Directorate, letter to author, 29 January 1997.
2. Worldwatch estimate based on sources cited in note 1.
3. Boden, Marland, and Andres, op. cit. note 1; Worldwatch estimate based on BP, op. cit. note 1.
4. Worldwatch estimate based on Boden, Marland, and Andres, op. cit note 1, on BP, op.

cit. note 1, and on Population Reference Bureau (PRB), "1996 World Population Data Sheet," wallchart (Washington, DC: June 1996).

5. Boden, Marland, and Andres, op. cit. note 1; BP, op. cit. note 1.
6. Boden, Marland, and Andres, op. cit. note 1; BP, op. cit. note 1.
7. Boden, Marland, and Andres, op. cit. note 1; BP, op. cit. note 1.
8. Boden, Marland, and Andres, op. cit. note 1; BP, op. cit. note 1.; PRB, op. cit. note 4.
9. Boden, Marland, and Andres, op. cit. note 1; BP, op. cit. note 1.
10. Boden, Marland, and Andres, op. cit. note 1; BP, op. cit. note 1; PRB, op. cit. note 4.
11. J.T. Houghton et al., eds., *Climate Change 1995: The Science of Climate Change*, Contribution of Working Group I to the Second Assessment Report of the Intergovernmental Panel on Climate Change (Cambridge, U.K.: Cambridge University Press, 1996).
12. Timothy Whorf and C.D. Keeling, Scripps Institution of Oceanography, La Jolla, CA, letter to author, 10 February 1997.
13. Houghton et al., op. cit. note 11.
14. Ibid.; R.T. Watson et al., eds., *Climate Change 1995: Impacts, Adaptations, and Mitigation*, Contribution of Working Group II to the Second Assessment Report of the Intergovernmental Panel on Climate Change (Cambridge, U.K.: Cambridge University Press, 1996).
15. Watson et al., op. cit. note 14.
16. *United Nations Framework Convention on Climate Change, Text* (Geneva: U.N. Environment Programme/World Meteorological Organization Information Unit on Climate Change, 1992).
17. "Industrial World is Unlikely to Meet Promised Carbon Cuts," *Nature*, 11 July 1996; DOE, EIA, *Annual Energy Outlook 1996* (Washington, DC: 1996); Climate Network Europe (CNE) and United States Climate Action Network (USCAN), *Independent NGO Evaluations of National Plans for Climate Change Mitigation: OECD Countries*, Fourth (Interim) Review, June 1996.
18. CNE and USCAN, op. cit. note 17.
19. Worldwatch estimate based on BP, op. cit. note 1, and on PRB, op. cit. note 4; International Energy Agency, *Climate Change Policy Initiatives, Selected Non-IEA Countries* (Paris: Organisation for Economic Co-operation and Development, 1996); Watson et al., op. cit. note 14.

20. International Energy Agency, *World Energy Outlook: 1996 Edition* (Paris: 1996).
21. "Nations Urged To Pass Laws On Emissions," *New York Times*, 19 July 1996.

SULFUR AND NITROGEN EMISSIONS UNCHANGED
(pages 60–61)

1. Dr. Jane Dignon, Lawrence Livermore National Laboratory, Livermore, CA, unpublished data series, letter to author, 22 January 1997; Sultan Hameed and Jane Dignon, "Global Emissions of Nitrogen and Sulfur Oxides in Fossil Fuel Combustion 1970-86," *Journal of the Air and Waste Management Association*, February 1992.
2. For a discussion of the methodology used to create this data set, see Sultan Hameed and Jane Dignon, "Changes in the Geographical Distributions of Global Emissions of NO$_X$ and SO$_X$ from Fossil Fuel Combustion between 1966 and 1980," *Atmospheric Environment* 22, no. 3 (1988), 441-49; Jane Dignon and Sultan Hameed, "Global Emissions of Nitrogen and Sulfur Oxides from 1860 to 1980," *JAPCA*, February 1989.
3. Dignon, op. cit. note 1.
4. Helen M. ApSimon and David Cowell, "The Benefits of Reduced Damage to Buildings from Abatement of Sulfur Dioxide Emissions," *Energy Policy*, July 1996.
5. Hameed and Dignon, op. cit. note 1.
6. Marc Levy, "European Acid Rain: The Power of Tote-Board Diplomacy," in Peter M. Haas et al., eds., *Institutions for the Earth* (Cambridge, MA: The MIT Press, 1993).
7. "Environmental Protection Agency's 1996 Auction of Sulfur Dioxide Emissions," *Coal Week*, 1 April 1996; Jeff Bailey, "Electric Utilities Are Overcomplying With Clean Air Act," *Wall Street Journal*, 15 November 1995.
8. Environmental Protection Agency, "EPA Report Shows Americans Breathing Cleaner Air While Economy Grows," press release, Washington, DC, 17 December 1996.
9. "EU Air Pollution Down in Early 1990s," *Reuter European Community Report*, 22 July 1996.
10. Per Elvingson, "Declaration Promises Unkept," *Acid News*, December 1996.
11. "Authorities Reveal 3 Million Deaths Linked to Illness from Urban Air Pollution," and "More than 70,000 Industrial Polluters Targeted for Closure under New Crackdown," both in *International Environment Reporter*, 30 October 1996; "China Adopts Effective Measures to Curb Pollution," *Xinhua News Agency*, 14 October 1996.
12. "India to Shut Polluters Near Taj," *Wall Street Journal*, 31 December 1996.
13. Peter M. Vitousek et al., "Human Alteration of the Global Nitrogen Cycle: Causes and Consequences" (draft), *Ecological Issues*, in press.
14. Robart Howarth, ed., "Nitrogen Cycling in the North Atlantic Ocean and Its Watersheds: Report of the International SCOPE Nitrogen Project," *Biogeochemistry*, October 1996 (special issue).
15. D.W. Schindler and S.E. Bayley, "The Biosphere as an Increasing Sink for Atmospheric Carbon: Estimates from Increasing Nitrogen Deposition," *Global Biogeochemistry Cycles*, vol. 7 (1993).
16. David Weldin and David Tilman, "Influence of Nitrogen Loading and Species Composition on the Carbon Balance of Grasslands," *Science*, 6 December 1996.

GLOBAL TEMPERATURE DOWN SLIGHTLY (pages 62–63)

1. James Hansen et al., Goddard Institute for Space Studies Surface Air Temperature Analyses, "Global Land-Ocean Temperature Index," as posted at < http://www.giss.nasa.gov/Data/GISTEMP >, viewed 14 January 1997; R. Monastersky, "1996: Year of Warmth and Weather Reversals," *Science News*, 18 January 1997.
2. Hansen et al., op. cit. note 1.
3. James Hansen et al., "1996 Temperature Observations," as posted at < http://www.giss.nasa.gov/Research/Observe/surftemp.html >, viewed 14 January 1997.
4. Ibid.
5. Hansen et al., op. cit. note 1; William K. Stevens, "Global Climate Stayed Warm in 1996, With Wet, Cold Regional Surprises," *New York Times*, 14 January 1997.
6. David Parker and Phil Jones, "Global Climate 1996—Not As Warm as 1995," U.K. Meteorological Office, Hadley Centre for Climate Prediction and Research and University of East Anglia Climatic Research

Unit, 15 January 1997.

7. Keith R. Briffa et al., "Unusual Twentieth-Century Warmth in a 1,000-Year Temperature Record from Siberia," *Nature*, 13 July 1995; Fred Pearce, "Lure of the Rings," *New Scientist*, 14 December 1996.

8. D.G. Vaughan and C.S.M. Doake, "Recent Atmospheric Warming and Retreat of Ice Shelves on the Antarctic Peninsula," *Nature*, 25 January 1996.

9. Wallace S. Broecker, "Chaotic Climate," *Scientific American*, November 1995; Jorge L. Sarmiento and Corinne Le Quere, "Oceanic Carbon Dioxide Uptake in a Model of Century-Scale Global Warming," *Science*, 22 November 1996; Fred Pearce, "Will a Sea Change Turn Up the Heat?" *New Scientist*, 30 November 1996.

10. Louis A. Codispoti, "Is the Ocean Losing Nitrate?" *Nature*, 31 August 1995.

11. R.T. Watson et al., eds., *Climate Change 1995: Impacts, Adaptations, and Mitigation*, Contribution of Working Group II to the Second Assessment Report of the Intergovernmental Panel on Climate Change (Cambridge, U.K.: Cambridge University Press, 1996); Chris Bright, "Tracking the Ecology of Climate Change," in Lester R. Brown et al., *State of the World 1997* (New York: W.W. Norton & Company, 1997).

12. David W. Schindler et al., "Consequences of Climate Warming and Lake Acidification for UV-B Penetration in North American Boreal Lakes," *Nature*, 22 February 1996; Fred Pearce, "Canadian Lakes Suffer Triple Blow," *New Scientist*, 24 February 1996.

13. George M. Woodwell and Fred T. MacKenzie, *Biotic Feedbacks in the Global Climatic System: Will the Warming Feed the Warming?* (New York: Oxford University Press, 1995).

14. J.T. Houghton et al., eds., *Climate Change 1995: The Science of Climate Change*, Contribution of Working Group I to the Second Assessment Report of the Intergovernmental Panel on Climate Change (Cambridge, U.K.: Cambridge University Press, 1996).

15. Ibid.

16. A.J. McMichael et al., eds., *Climate Change and Human Health*, prepared by a Task Group on behalf of the World Health Organization (WHO), World Meteorological Organization, and United Nations Environment Programme (Geneva: WHO, 1996).

17. B.D. Santer et al., "A Search for Human Influences on the Thermal Structure of the Atmosphere," and Neville Nichols, "An Incriminating Fingerprint," both in *Nature*, 4 July 1996; Richard A. Kerr, "Sky-High Findings Drop New Hints of Greenhouse Warming," *Science*, 5 July 1996; Figure 2 from The Hadley Centre for Climate Prediction and Research, *Modeling Climate Change 1860–2050* (Bracknell, U.K.: The Meteorological Office, February 1995).

18. J. Hansen et al., "A Pinatubo Climate Modeling Investigation," in G. Fiocco, D. Fua, and G. Visconti, eds., *Global Environmental Change* (Berlin: Springer-Verlag, 1996).

19. Hansen et al., op. cit. note 3.

WORLD ECONOMY EXPANDS FASTER (pages 66–67)

1. International Monetary Fund (IMF), *World Economic Outlook, October 1996* (Washington, DC: 1996).

2. Ibid.

3. Ibid.

4. Ibid.

5. Ibid.

6. Ibid.

7. Ibid.

8. Andrew Pollack, "The Question Facing Japan: Can Its Vibrant Engine Ever Be Restarted?" *New York Times*, 2 January 1997.

9. IMF, op. cit. note 1.

10. Ibid.

11. Ibid.

12. Ibid.

13. Ibid.

14. "Vietnamese Economy At A Crossroads," *Journal of Commerce*, 21 June 1996.

15. IMF, op. cit. note 1.

16. Ibid.

17. Ibid.

18. Ibid.

19. Ibid.

20. "Mexican Output Rises 7.4%, Eclipsing Economic Forecasts," *New York Times*, 22 November 1996.

21. IMF, op. cit. note 1.

22. Ibid.

23. Ibid.

24. Ibid.

25. Ibid.

26. Ibid.

27. Ibid.

28. Ibid.

ROUNDWOOD PRODUCTION
RISES AGAIN (pages 68–69)

1. U.N. Food and Agriculture Organization (FAO), *Forest Products Yearbooks* (Rome: various years); 1995 from Mafa Chipeta, FAO, Rome, discussion with author, 30 December 1996.
2. Chipeta, op. cit. note 1.
3. Nigel Dudley, Jean-Paul Jeanrenaud, and Francis Sullivan, *Bad Harvest? The Timber Trade and the Degradation of the World's Forests* (London: Earthscan Publications, 1995).
4. Chipeta, op. cit. note 1.
5. Ibid.
6. German Bundestag, ed., *Protecting the Tropical Forests: A High Priority International Task* (Bonn: Bonner Universitats-Buchdruckerei, 1990).
7. FAO, *Forest Products Yearbook, 1983–94* (Rome: 1996).
8. German Bundestag, op. cit. note 6.
9. FAO, op. cit. note 7.
10. Chipeta, op. cit. note 1.
11. James McIntire, ed., *The New Eco-Order: Economic and Ecological Linkages of the World's Temperate and Boreal Forest Resources* (Seattle, WA: Northwest Policy Center, University of Washington, 1995).
12. Nigel Dudley and Sue Stolton, "Pulp Fact: The Environmental and Social Impacts of the Pulp and Paper Industry," World Wide Fund for Nature, < http://www.panda.org/tda/forest/ contents.htm >, Gland, Switzerland, updated 8 May 1996.
13. Ibid.
14. Ibid.
15. Ibid.
16. Dudley, Jeanrenaud, and Sullivan, op. cit. note 3; FAO, *Forestry Statistics: Today and Tomorrow* (Rome: 1995).
17. International Institute for Environment and Development, for World Business Council for Sustainable Development (WBCSD), *A Changing Future for Paper* (Geneva: WBCSD, 1996).
18. Dudley and Stolton, op. cit. note 12
19. Jamison Ervin, Forest Stewardship Council-US Initiative, Waterbury, VT, discussion with author, 30 December 1996; see also Cheri Sugal, "Labeling Wood: How Timber Certification May Reduce Deforestation," *World Watch*, September/October 1996.

20. FAO, op. cit. note 7.
21. McIntire, op. cit. note 11; Dudley, Jeanrenaud, and Sullivan, op. cit. note 3.

STORM DAMAGES SET RECORD
(pages 70–71)

1. Gerhard A. Berz, Muchener Ruckver-sicherungs-Gesellschaff, press release (Munich, Germany: 23 December 1996).
2. Ibid.
3. Ibid.
4. Craig S. Smith, "Rains Cause Lethal Flooding in China, Forcing Evacuations and Ruining Crops," *Wall Street Journal*, 8 August 1996.
5. Berz, op. cit. note 1.
6. Angus MacSwan, "'96 Hurricanes Claimed Heavy Toll," *Boston Globe*, 29 November 1996.
7. Berz, op. cit. note 1.
8. "Recent California Flooding Will Cost Taxpayers a Pretty Penny," *San Jose Mercury News*, 15 January 1997.
9. Karl cited in Richard Cole, "Data Suggest Long-Term Change Toward Heavy Weather Worldwide," *Washington Post*, 21 January 1997.
10. Ibid.
11. Berz, op. cit. note 1.
12. Ibid.
13. Ibid.
14. Emmeline Legerwood, "Cutting the Cost of a Catastrophe," *Financial Times*, 28 August 1996.
15. UNEP Insurance Initiative, *Position Paper on Climate Change* (Geneva: U.N. Environment Programme, 9 July 1996).

AUTOMOBILE FLEET EXPANDS
(pages 74–75)

1. Worldwatch estimate based on American Automobile Manufacturers Association (AAMA), *World Motor Vehicle Data*, 1996 ed. (Detroit, MI: 1996), on AAMA, *Motor Vehicle Facts and Figures 1996* (Detroit, MI: 1996), and on DRI/McGraw-Hill, *World Car Industry Forecast Report* (London: November 1996).
2. AAMA, *World Motor Vehicle Data*, op. cit. note 1.
3. Ibid.; population data from U.S. Bureau of the Census, *International Data Base*, electronic database, Suitland, MD, updated 15 May 1996.

4. Worldwatch estimate based on DRI/McGraw-Hill, op. cit. note 1; John Griffiths, "Car Sales Leap in Western Europe," *Financial Times*, 14 November 1996.
5. DRI/McGraw-Hill, op. cit. note 1.
6. Ibid.; Gregory Platt, "Foreign Automakers Flock to Poland," *Christian Science Monitor*, 18 January 1996.
7. DRI/McGraw-Hill, op. cit. note 1.
8. Ibid.
9. Ibid.; "Latin America's Car Industry: Revving Up," *The Economist*, 27 April 1996; John Griffiths, "Emerging Nations Power World Vehicle Sales," *Financial Times*, 13 May 1996.
10. DRI/McGraw-Hill, op. cit. note 1.
11. Ibid.; John Griffiths, "Uncertain Future for Asian Cars," *Financial Times*, 28 November 1996.
12. DRI/McGraw-Hill, op. cit. note 1.
13. Ibid.
14. Griffiths, op. cit. note 11; Tony Walker, "Rocky Road Lies Ahead for China's Car Industry," *Financial Times*, 29 October 1996; "China Faces Sluggish 1997 Auto Market Sales," *Journal of Commerce*, 17 December 1996.
15. Manjeet Kripalani, "A Traffic Jam of Auto Makers," *Business Week*, 5 August 1996; DRI/McGraw-Hill, op. cit. note 1.
16. Paul A. Eisenstein, "Korea, Poland Typify Global Automotive Boom," *Christian Science Monitor*, 18 January 1996.
17. Andrew Pollack, "The Race for a Winning Asia Car," *New York Times*, 6 June 1996; Nopporn Wang-Anan, "Ford Studies Development of Car for Asia," *Wall Street Journal*, 25 November 1996.
18. Michael Hsu, "China: The Glittering Prize," *Journal of Commerce*, 15 August 1996; Chang Weimin. "Race On for Car Output," *China Daily*, 4 March 1996.
19. John Griffiths, "Car Numbers to Double and Threaten the Environment," *Financial Times*, 15 February 1996.
20. James J. MacKenzie, Roger C. Dower, and Donald D.T. Chen, *The Going Rate: What It Really Costs to Drive* (Washington, DC: World Resources Institute (WRI), 1992).
21. WRI et al., *World Resources 1996–97* (New York: Oxford University Press, 1996).
22. Molly Moore, "Mexico City Gasping In Quest of Fair Air," *Washington Post*, 25 November 1996; WRI et al., op. cit. note 21.
23. James J. MacKenzie and Michael P. Walsh, *Driving Forces* (Washington, DC: WRI, 1990);

Natural Resources Defense Council, *Green Auto Racing: National Efforts and International Cooperation to Promote Advanced Cars and Fuels* (Washington, DC: July 1996).
24. WRI et al., op. cit. note 21.
25. Iain Carson, "Taming the Beast," *The Economist*, 22 June 1996; Jessica Mathews, "Cars, Cars, Cars," *Washington Post*, 30 September 1996.
26. WRI et al., op. cit. note 21; Ted Badarcke, "Bangkok May Ban New Cars Until 2001," *Financial Times*, 8 July 1996.
27. Dean E. Murphy, "A Bicycle Built for 5.2 Million?" *Los Angeles Times*, 31 May 1996; WRI et al., op. cit. note 21.
28. Michael Replogle, *Non-Motorized Vehicles in Asian Cities*, Technical Paper Number 162 (Washington, DC: World Bank, 1992); International Institute for Energy Conservation, *Moving Toward Integrated Transport Planning: Energy, Environment and Mobility in Four Asian Cities* (Washington, DC: 1993).
29. Mathews, op. cit. note 25; Ted Badarcke, "Nightmare of Bangkok's Jams," *Financial Times*, 3 June 1996.
30. World Bank, *Sustainable Transport: Priorities for Policy Reform* (Washington, DC: 1996).

BICYCLE PRODUCTION STILL RISING (pages 76–77)

1. "World Market Report," *1997 Interbike Directory* (Newport Beach, CA: Primedia, Inc. 1997); United Nations, *Industrial Commodities Statistics Yearbook 1994* (New York: United Nations, 1996). Total production differs from that in earlier editions of *Vital Signs* due to the use of the United Nations data.
2. "World Market Report," op. cit. note 1; United Nations, op. cit. note 1.
3. Marc Sani, "Big Inventory Mars Strong Sales in 1995," *Bicycle Retailer and Industry News*, 1 February 1996; Michael Gamstetter, "Bicycle Glut Slows Taiwan Sales Growth," *Bicycle Retailer and Industry News*, 15 July 1996.
4. "World Market Report," op. cit. note 1.
5. Ibid.
6. Ibid.
7. Ibid.
8. "World Market Report," *Interbike Directory*, various years.
9. Ibid.
10. Ibid.

11. Ibid.

12. John H. Cushman, "For Bicyclists Who Welcome a Little Help (Mostly Uphill)," *New York Times*, 8 August 1996.

13. "New Yamaha Electric Bike Ups Performance," *Bicycle Retailer and Industry News*, 1 January 1997.

14. "Progress Towards a $15 Dollar Bicycle," *Sustainable Transport*, Summer 1996.

15. Ibid.; Deike Peters, "Bikeways Come to Lima's Mean Streets," *Sustainable Transport*, Winter 1997.

16. "Living Dangerously," *Sustainable Transport*, Summer 1996.

17. Federal Highway Administration (FHA), "Reasons Why Bicycling and Walking Are and Are Not Being Used More Extensively as Travel Modes," Case Study Number 1, *National Bicycling and Walking Study* (Washington, DC: U.S. Department of Transportation, 1994).

18. "New York: Safety Costs Put Bike Messengers Out of Business," *Sustainable Transport*, Summer 1996.

19. Sonia Schlossman, "On Your Bike In Copenhagen," *Blue Wings* (magazine of Finnair), February–March 1996.

20. Ibid.

21. Copenhagen from Urs Heierli, *Environmental Limits to Motorisation: Non-Motorized Transport in Developed and Developing Countries* (St. Gallen, Switzerland: Swiss Centre for Development Cooperation in Technology and Management, 1993); FHA, op. cit. note 17.

22. Heierli, op. cit. note 21.

23. Deike Peters, "Bikeways Come to Lima's Mean Streets," *Sustainable Transport*, Winter 1997.

24. Ibid.

25. "Transport Now Consumes More Energy Than Industry," *Sustainable Transport*, Winter 1997; "EU Sees Growing Role for the Bicycle," *Bicycle Retailer and Industry News*, 15 March 1996.

26. "Government Sets Target to Quadruple Bicycle Use," *ENDS Report 258*, July 1996.

POPULATION INCREASE SLOWS SLIGHTLY (pages 80–81)

1. U.S. Bureau of the Census, *International Data Base*, Suitland, MD, 15 May 1996.

2. U.S. Bureau of the Census, *World Population Profile: 1996* (Washington, DC: U.S. Government Printing Office, July 1996).

3. Bureau of the Census, op. cit. note 1.

4. Ibid.

5. Peter Johnson, U.S. Bureau of the Census, discussion with author, 7 January 1997; Barbara Crossette, "World is Less Crowded Than Expected, The UN Reports," *New York Times*, 11 November 1996.

6. Population Reference Bureau (PRB), "1996 World Population Data Sheet," wallchart (Washington, DC: June 1996); Bureau of the Census, op. cit. note 2; Johnson, op. cit. note 5.

7. United Nations, *World Population Prospects: The 1994 Revision* (New York: 1995).

8. Bureau of the Census, op. cit. note 1; Bureau of the Census op. cit. note 2; Johnson, op. cit. note 5.

9. Bureau of the Census, op. cit note 1.

10. Ibid.; Toni Nelson, "Russia's Population Sink," *World Watch*, January/February 1996.

11. Bureau of the Census, op. cit. note 1.

12. Bureau of the Census, op. cit. note 2; PRB, op. cit. note 6.

13. Bureau of the Census, op. cit. note 2.

14. Ibid.

15. Ibid.; Population Action International (PAI), *Why Population Matters: 1996* (Washington, DC: 1996).

16. Bureau of the Census, op. cit. note 2.

17. Bureau of the Census, op. cit. note 1.

18. Bureau of the Census, op. cit. note 2; Judy Mann, "Time to Reverse a Family Planning Fiasco," *Washington Post*, 24 January 1997.

19. PAI, op. cit. note 15; Mann, op. cit. note 18.

20. Anton de Jong, Netherlands Permanent Mission to the UN, presentation at the PAI seminar "UNFPA: Meeting the Challenge of the 21st Century," Washington, DC, 24 January 1997; Christopher Flavin, "The Legacy of Rio," in Lester R. Brown et al., *State of the World 1997* (New York: W.W. Norton & Company, 1997); PAI, *1995 Annual Report* (Washington, DC: 1996).

21. Kerstin Trone, Deputy Executive Director, United Nations Population Fund, presentation at the PAI seminar "UNFPA: Meeting the Challenge of the 21st Century," Washington, DC, 24 January 1997; Crossette, op. cit. note 5.

22. "Statement By Dr. Rafiq Zakaria," Member of the Indian Delegation to the UN Commission on Population and Development, as presented in the Report of the Secretary-General in the 29th Session, < gopher://gopher,undp.org: 70/00/ungophers/popin/unpopcom/29thses/ statements/india2 >, New York, posted 28

February 1996.

REFUGEE POPULATION HAS RARE DECLINE (pages 82–83)

1. Food and Statistical Unit, Division of Programmes and Operational Support, *Populations of Concern to UNHCR: A Statistical Overview* (Geneva: U.N. High Commissioner For Refugees (UNHCR), July 1996); UNHCR, "UNHCR by Numbers," data sheet (Geneva: 1996).
2. Historical data from UNHCR, *The State of the World's Refugees 1995* (New York: Oxford University Press, 1995); recent data from UNHCR, "UNHCR at a Glance," data sheet, Geneva, November 1995.
3. UNHCR, *The State of the World's Refugees 1995*, op. cit. note 2; UNHCR, "UNHCR at a Glance," op. cit. note 2.
4. Hal Kane, *The Hour of Departure: Forces That Create Refugees and Migrants*, Worldwatch Paper 125 (Washington, DC: Worldwatch Institute, June 1995).
5. UNHCR, op. cit. note 1.
6. Ibid.
7. Ibid.
8. Ibid.
9. Ibid.
10. U.S. Committee for Refugees, *World Refugee Survey 1996* (Washington, DC: 1996).
11. UNHCR, op. cit. note 1.
12. Ibid.
13. Ibid.
14. Ibid.
15. Michael Renner, *Fighting for Survival: Environmental Decline, Social Conflict, and the New Age of Insecurity* (New York: W.W. Norton & Company, 1996).
16. Ibid.
17. UNHCR, op. cit. note 1.
18. Ibid.
19. Ibid.; UNHCR, *The State of the World's Refugees*, op. cit. note 2.
20. UNHCR, op. cit. note 1; UNHCR, *The State of the World's Refugees*, op. cit. note 2.
21. UNHCR, op. cit. note 1.
22. International Red Cross and Red Crescent, *World Disasters Report* (Geneva: 1995).
23. Ibid.

HIV/AIDS PANDEMIC BROADENS REACH (pages 84–85)

1. Daniel Tarantola, Global AIDS Policy Coalition (GAPC), Harvard School of Public Health, Cambridge, MA, discussion with author, 24 January 1997. Data for 1996 are also available from UNAIDS, but no historic series is available. GAPC's and UNAIDS estimates differ for two main reasons: GAPC estimates assume a shorter period between HIV infection and onset of full-blown AIDS for most developing regions, thus resulting in more cases of and more deaths from AIDS; GAPC estimates a larger number of infections in Southeast Asia, particularly in India.
2. Tarantola, op. cit. note 1.
3. Ibid.
4. Jonathan Mann and Daniel Tarantola, eds., *AIDS in the World II* (New York: Oxford University Press, 1996).
5. U.S. Bureau of the Census, *World Population Profile: 1996* (Washington, DC: U.S. Government Printing Office, June 1996); UNAIDS, "HIV/AIDS: The Global Epidemic, December 1996," fact sheet (Geneva: 28 November 1996).
6. Mann and Tarantola, op. cit. note 4.
7. John F. Burns, "Denial and Taboo Blind India to the Horror of Its AIDS Scourge," *New York Times*, 22 September 1996.
8. Lee Hockstader, "With Plague's Fury, HIV Spreads in Belarus Town," *Wall Street Journal*, 26 November 1996.
9. Ibid.; UNAIDS, "UNAIDS World AIDS Day Report Documents HIV Threat to New Global Populations," press release (London: 28 November 1996).
10. Lawrence K. Altman, "UN Reports 3 Million New H.I.V. Cases Worldwide for '96," *New York Times*, 28 November 1996.
11. UNAIDS, op. cit. note 5; AIDSCAP/Family Health International, Harvard School of Public Health, and UNAIDS, "The Status and Trends of The Global HIV/AIDS Pandemic—Symposium Final Report" (Vancouver: 5–6 July 1996); "A Violation of Human Rights?" *Earth Times: AIDS Report*, 15–30 July 1996.
12. UNAIDS, op. cit. note 5.
13. Ibid.
14. "Not One Epidemic But Several," *Earth Times: AIDS Report*, 15–30 July 1996; UNAIDS, op. cit. note 5.
15. AIDSCAP/Family Health International, Harvard School of Public Health, and UNAIDS, op. cit.

note 11.

16. Tarantola, op. cit. note 1.
17. Susan Okie, "Lessons From Africa in AIDS Prevention," *Washington Post*, 16 December 1996; AIDSCAP/Family Health International, Harvard School of Public Health, and UNAIDS, op. cit. note 11.
18. Joanne Kenen, "Needle Exchange Program May Slow AIDS," *Detroit News*, 20 September 1995; Center for AIDS Prevention Studies at The University of California, "Does Needle Exchange Work?" < http://www. epibiostat. ucsf.edu/capsweb/needletext.html >, San Francisco, viewed 12 February 1997.
19. Andrew Purvis, "The Global Epidemic," *Time*, 30 December 1996–6 January 1997.

U.N. PEACEKEEPING DECLINES SHARPLY (pages 88–89)

1. Figures for 1993–96 from "Peace-Keeping Operations Expenditures (All Missions)," < http://www.un.org/Depts/DPKO/yir96/allexp.j pg >, viewed 18 March 1997; 1986–92 figures from Luisa Anzola, U.N. Department of Peacekeeping Operations, New York, discussion with author, 20 December 1995; pre-1986 expenditures calculated from Joseph Preston Baratta, *International Peacekeeping: History and Strengthening* (Washington, DC: Center for U.N. Reform Education, 1989), and from U.N. Department of Public Information (UNDPI), *United Nations Peace-Keeping* (New York: 1993).
2. Michael Renner, "Peacekeeping Expenditures Level Off," in Lester R. Brown, Christopher Flavin, and Hal Kane, *Vital Signs 1996* (New York: W.W. Norton & Company, 1996).
3. Figure for 1995 from ibid.; 1996 from Mikhail Seliankin, UNDPI, New York, discussion with author, 30 November 1996.
4. United Nations, "Completed Peace-keeping Operations," < http://www.un.org/Depts/DPKO/ p_miss.htm >, New York, viewed 6 January 1997.
5. Trevor Findlay, "Armed Conflict Prevention, Management and Resolution," in Stockholm International Peace Research Institute (SIPRI), *SIPRI Yearbook 1996. Armaments, Disarmament and International Security* (New York: Oxford University Press, 1996). The Support Mission in Haiti (UNSMIH) was kicked off in July 1996, but it is a direct successor to the previous

Mission in Haiti (UNMIH); likewise, the Observer Mission in El Salvador (ONUSAL) gave way in 1995 to a follow-up mission, MINUSAL.
6. United Nations, "Angola—UNAVEM III," < http://www.un.org/Depts/DPKO/Missions/un avem3.htm >, viewed 6 January 1997.
7. Elaine Sciolino, "Accord Reached to End the War in Bosnia; Clinton Pledges U.S. Troops to Keep Peace," *New York Times*, 22 November 1995.
8. International Institute for Strategic Studies, *The Military Balance 1996/97* (London: Oxford University Press, October 1996); IFOR Web site maintained by NATO, < http://www.nato. int/ifor/ >, viewed 6 January 1997; Craig R. Whitney, "NATO Clears Smaller, U.S.-Led Force to Extend Bosnia Mission," *New York Times*, 18 December 1996.
9. United Nations, "Current Peace-keeping Operations," < http://www.un.org/Depts/DPKO/ c_miss.htm >, New York, viewed 6 January 1997.
10. Cost in 1996 from Seliankin, op. cit. note 3; 1995 from Renner, op. cit. note 2.
11. Some of the conflicts required a succession of missions; in the former Yugoslavia, for instance, six operations were created between 1992 and 1996; calculated from "Peace-Keeping Operations," op. cit. note 1.
12. Number of people from UNDPI, "Background Note: United Nations Peace-Keeping Operations" (New York: 1 March 1996); number of countries from UNDPI, "U.N. Peace-Keeping: Some Questions and Answers" (New York: September 1996).
13. UNDPI, "U.N. Peace-Keeping," op. cit note 12.
14. United Nations, Office of the Spokesman of the Secretary-General, "Outstanding Contributions to the Regular Budget, International Tribunals and Peace-Keeping Operations" (New York: 7 January 1997).
15. Ibid.
16. Ibid.
17. See Harry Dunphy, "Clinton Plans to Pay U.N. Debt," Associated Press, 9 January 1997, on CompuServe's Time Magazine site, for most recent proposal.
18. Steven Lee Myers, "On Capitol Hill, Praise for U.N. Leader But No Firm Word on Aid," *New York Times*, 25 January 1997.
19. United Nations, "The Lessons Learned Unit," < http://www.un.org/Depts/dpko/llu2.htm >, viewed 6 January 1997.

20. Michael Renner, *Remaking U.N. Peacekeeping: U.S. Policy and Real Reform*, ECD Briefing Paper No. 17 (Washington, DC: National Commission for Economic Conversion and Disarmament, November 1995); United Nations, "Ghana Joins Stand-By Arrangements for More Rapid Deployment of Peace-Keeping Troops," press release (New York: 29 May 1996); United Nations, "Malaysia Joins Stand-By Arrangements for More Rapid Deployment of Peace-Keeping Troops," press release (New York: 6 September 1996).

21. United Nations, "Haiti—UNSMIH," < http://www.un.org/Depts/DPKO/Missions/unsmih.htm >, viewed 15 December 1996.

22. Paul Lewis, "China Lifts U.N. Veto on Guatemala Monitors," *New York Times*, 21 January 1997; Sierra Leone from U.N. Daily Highlights, press release, 18 December 1996, as posted at < http://www.un.org/cgi-bin/dh.pl >.

23. "Central African Republic in 3-Way Peace Accord," *New York Times*, 26 January 1997.

ARMED FORCES CONTINUE DEMOBILIZATIONS (pages 90–91)

1. U.S. Arms Control and Disarmament Agency (ACDA), *World Military Expenditures and Arms Transfers (WMEAT)*, electronic database, Washington, DC, as provided by Daniel Gallik, WMEAT Editor, ACDA, January 1997.

2. Ibid.

3. ACDA, *World Military Expenditures and Arms Transfers 1995* (Washington, DC: U.S. Government Printing Office, April 1996).

4. International Institute for Strategic Studies (IISS), *The Military Balance 1996/97* (London: Oxford University Press, 1996).

5. Ibid.

6. First ratio is from ACDA, op. cit. note 3; second ratio is a Worldwatch calculation based on IISS, op. cit. note 4.

7. ACDA, op. cit. note 3.

8. IISS, op. cit. note 4.

9. Total outside own countries calculated from IISS, op. cit. note 4; soldiers in peacekeeping is a Worldwatch estimate based on Mikhail Seliankin, U.N. Department of Public Information, New York, discussion with author, 30 November 1996, on IISS, op. cit. note 4, and on Trevor Findlay, "Armed Conflict Prevention, Management and Resolution," in Stockholm International Peace Research Institute (SIPRI), *SIPRI Yearbook 1996. Armaments, Disarmament and International Security* (New York: Oxford University Press, 1996).

10. U.S. Department of Defense, Public Affairs Office, discussion with author, 23 December 1996. Figure current as of September 1996.

11. Current Russian figure calculated from IISS, op. cit. note 4; Soviet figure from Bonn International Center for Conversion (BICC), *Conversion Survey 1996. Global Disarmament, Demilitarization and Demobilization* (New York: Oxford University Press, 1996).

12. Calculated from IISS, op. cit. note 4.

13. Among the countries with large armed opposition groups, IISS has no data for Algeria, Burundi, Chad, India, Mexico, Pakistan, Russia, Senegal, and Zaire. SIPRI puts armed opposition forces in Chechnya at 12,000-20,000; Margareta Sollenberg and Peter Wallensteen, "Major Armed Conflicts," in SIPRI, op. cit. note 9.

14. Data are for 1994; ACDA, op. cit. note 3.

15. ACDA, op. cit. note 3.

16. IISS, op. cit. note 4.

17. Costa Rica from Paul George et al., "Military Expenditure," in SIPRI, op. cit. note 9; Joaquín Tacsan, "Report on Projects and Activities of the Center for Peace and Reconciliation," in Arias Foundation, *Arias Foundation for Peace and Human Progress Performance Report 1988–1996* (San José, Costa Rica: 1996).

18. BICC, op. cit. note 11.

19. Ibid.

20. U.S. Department of Defense, Office of the Undersecretary of Defense (Comptroller), *National Defense Budget Estimates for FY 1996* (Washington, DC: National Technical Information Service, March 1995).

21. BICC, op. cit. note 11; ACDA, op. cit. note 3.

22. BICC, op. cit. note 11; Howard W. French, "Liberian Militias Lay Down Arms and Raise Hopes," *New York Times*, 27 January 1997.

23. *International Security Digest*, October 1996.

24. BICC, op. cit. note 11.

25. Ibid.; World Bank, *Demobilization and Reintegration of Military Personnel in Africa: The Evidence from Seven Country Case Studies*, Africa Regional Series Discussion Paper IDP-130 (Washington, DC: October 1993).

FOREST LOSS CONTINUES
(pages 96–97)

1. U.N. Food and Agriculture Organization (FAO), *State of the World's Forests 1997* (Rome: 1997).
2. Preagricultural calculations based on Table 1 in R.A. Houghton et al., "Changes in the Carbon Content of Terrestrial Biota and Soils between 1860 and 1980: A Net Release of CO_2 to the Atmosphere," *Ecological Monographs*, September 1983.
3. FAO, op. cit. note 1.
4. Ibid.
5. Ibid.
6. Ibid.
7. Ibid.
8. Ibid.
9. FAO, *Forest Resources Assessment: Global Synthesis*, Forestry Paper 124 (Rome: 1995).
10. Ibid.
11. FAO, op. cit. note 1.
12. Michael Williams, "Forests," in B.L. Turner II et al., eds., *The Earth as Transformed by Human Action: Global and Regional Changes in the Biosphere over the Past 300 Years* (Cambridge, U.K.: Cambridge University Press, 1990).
13. Forest areas from FAO, op. cit. note 1; percentage of U.S. area logged from Dominic DellaSala, Forest Conservation Director, World Wildlife Fund-US, presentation at Worldwatch Institute, Washington, DC, 4 November 1996.
14. L.R. Oldeman, R.T.A. Hakkeling, and W.G. Sombroek. *World Map of the Status of Human-Induced Soil Degradation: An Explanatory Note* (Wageningen, the Netherlands, and Nairobi: International Soil Reference and Information Centre and U.N. Environment Programme, 1991).
15. Paul R. Ehrlich and Anne H. Ehrlich, "The Value of Biodiversity," *Ambio*, May 1992.
16. Deutscher Bundestag, *Protecting the Tropical Forests: A High Priority International Task* (Bonn: 1990).
17. FAO, op. cite. note 9.
18. Caroline Wheat, "Biodiversity: What Is It, Where Is It, and Why Should We Care?" *Calypso Log*, February 1994.
19. Norman R. Farnsworth, "Screening Plants for New Medicines," in E.O. Wilson, ed., *Biodiversity* (Washington, DC: National Academy Press, 1988).
20. T.C. Whitmore and J.A. Sayer, *Tropical Deforestation and Species Extinction* (London: Chapman & Hall, 1992).

21. Deutscher Bundestag, op. cite. note 16.
22. Gareth Porter, "Managing Renewable Resources in Southeast Asia: The Problem of Deforestation," in Young Kim, ed., *The Southeast Asian Economic Miracle* (New Brunswick, NJ: Transaction Publishers, 1995).
23. Ibid.
24. Ibid.
25. World Wide Fund for Nature-UK (WWF-UK) and World Conservation Monitoring Centre, "World Forest Map," WWF International, <http://www.panda.org>, Gland, Switzerland, viewed 20 February 1997.
26. Don Gilmore, "The Global Forest Debate—Where Are We Going?" *Arborvitae: The IUCN/WWF Forest Conservation Newsletter*, November 1996.

ECOSYSTEM CONVERSION SPREADS (pages 98–99)

1. Table 1 calculated from U.N. Food and Agriculture Organization (FAO), "FAOSTAT-PC," electronic database, Rome, 1995, from FAO, *Forest Resources Assessment 1990: Global Synthesis*, FAO Forestry Paper 124 (Rome: 1995), and from World Resources Institute (WRI) et al., *World Resources 1996–97* (New York: Oxford University Press, 1996).
2. Worldwatch estimate, calculated from FAO, "FAOSTAT-PC," op. cit. note 1, from FAO, Forestry Paper 124, op. cit. note 1, and from WRI et al., op. cit. note 1.
3. Worldwatch estimate, calculated from FAO, "FAOSTAT-PC," op. cit. note 1, from FAO, Forestry Paper 124, op. cit. note 1, and from WRI et al., op. cit. note 1.
4. Worldwatch estimate, calculated from FAO, "FAOSTAT-PC," op. cit. note 1, from FAO, Forestry Paper 124, op. cit. note 1, and from WRI et al., op. cit. note 1.
5. Worldwatch estimate, calculated from FAO, "FAOSTAT-PC," op. cit. note 1, from FAO, Forestry Paper 124, op. cit. note 1, and from WRI et al., op. cit. note 1.
6. Europe and Asia from Michael Moser, Crawford Prentice, and Scott Frazier, "A Global Overview of Wetland Loss and Degradation," prepared for the Conference of the Parties to the Ramsar Convention (Slimbridge, U.K.: Wetlands International, March 1996); U.S. estimate from T.E. Dahl, *Wetland Losses in the United States,*

1780's to 1980's (Washington, DC: Department of the Interior, Fish and Wildlife Service, 1990).

7. Philippines estimate from Moser, Prentice, and Frazier, op. cit. note 6.

8. Solon Barraclough and Andrea Finger-Stich, *Some Ecological and Social Implications of Commercial Shrimp Farming in Asia*, Discussion Paper 74 (Geneva: United Nations Research Institute for Social Development, March 1996).

9. Elliot Norse, President, Marine Conservation Biology Institute, discussion with author, 2 February 1997.

10. Elliot Norse, "Bottom Trawling: The Unseen Worldwide Plowing of the Seabed," *New England Biolabs Transcript*, January 1997.

11. FAO, *State of the World's Forests 1997* (Rome: 1997).

12. Worldwatch estimate, calculated from FAO,"FAOSTAT-PC," op. cit. note 1, from FAO, Forestry Paper 124, op. cit. note 1, and from WRI et al., op. cit. note 1.

PRIMATE DIVERSITY DWINDLING WORLDWIDE
(pages 100–01)

1. World Conservation Union–IUCN, *1996 IUCN Red List of Threatened Animals* (Gland, Switzerland: 1996).

2. Ibid.

3. Worldwatch rough calculation based on IUCN estimates of nonhuman primate numbers, compared with a 1996 human population of 5.8 billion, from Population Reference Bureau, "1996 World Population Data Sheet," wallchart (Washington, DC: June 1996).

4. Ronald M. Nowak, *Walker's Mammals of the World, Fifth Edition, Vol. 1* (Baltimore, MD: Johns Hopkins University Press, 1991).

5. Species total from Anthony B. Rylands, Russell A. Mittermeier, and Ernesto Rodriguez Luna, "A Species List for the New World Primates (Platyrrhini): Distribution by Country, Endemism, and Conservation Status According to the Mace-Land System," *Neotropical Primates*, September 1995; new species record for 1996 from Rick Weiss, "Brazil Reveals Another New Primate Species," *Washington Post*, 20 June 1996.

6. Indonesia figure from A.A. Eudey, *Action Plan for Asian Primate Conservation: 1987–1991* (Gland, Switzerland: IUCN, 1986); Zaire figure from John F. Oates, *African Primates* (Gland, Switzerland: IUCN, 1996); Russell A. Mittermeier et al., *Lemurs of Madagascar: An Action Plan for their Conservation* (Gland, Switzerland: IUCN, 1992).

7. Mittermeier et al., op. cit. note 6.

8. IUCN, op. cit. note 1.

9. Russell A. Mittermeier, Warren G. Kinzey, and Roderic B. Mast, "Neotropical Primate Conservation," in P. Arambulo III et al., eds., *Primates of the Americas: Strategies for Conservation and Sustained Use in Biomedical Research* (Columbus, OH: Battelle Press, 1990).

10. Rylands, Mittermeier, and Rodriguez Luna, op. cit. note 5.

11. Frank Kuznik, "How to Be an Orangutan," *International Wildlife*, January/February 1997.

12. IUCN, op. cit. note 1.

13. John E. Fa et al., "Impacts of Market Hunting on Mammal Species in Equatorial Guinea," *Conservation Biology*, October 1995.

14. David S. Wilkie, John G. Sidle, and Georges C. Boundzanga, "Mechanized Logging, Market Hunting, and a Bank Loan in Congo," *Conservation Biology*, December 1992.

15. Ibid.

16. Ibid.

17. Ardith A. Eudey, "Captive Gibbons in Thailand and the Option of Reintroduction to the Wild," *Primate Conservation*, 1991–1992.

18. "Wild-Primate Exports Stop," *HSUS News* (Humane Society of the United States), Spring 1994.

19. J.R. Held and T.L. Wofle, "Imports: Current Trends and Usage," *American Journal of Primatology*, 34 (1994), 85–96.

20. Ibid.; R.A. Whitney, Jr., "The Conservation of Nonhuman Primates and Its Importance to Public Health," in Arambulo et al., op. cit. note 9.

21. Figure of 50 percent from Held and Wofle, op. cit. note 19; 80 percent estimate from "Wild-Primate Exports Stop," op. cit. note 18.

22. Zena Tooze, "Update on Sclater's Guenon *Cercopithecus sclateri* in Southern Nigeria," *African Primates*, December 1995.

23. Dionizio M. Pessamilio, "Revegetation of Deforested Areas in the Poco das Antas Biological Reserve, Rio de Janeiro," *Neotropical Primates*, December 1994.

24. Paul F. Salopek, "Gorillas and Humans: An Uneasy Truce," *National Geographic*, October 1995.

25. Gillian Haggerty, "Conservation Heroes: Gorilla

Guards Win J. Paul Getty Award," *WWF News* (World Wide Fund for Nature), Summer 1996.

OZONE RESPONSE ACCELERATES
(pages 102–03)

1. Richard Elliot Benedick, *Ozone Diplomacy* (Cambridge, MA: Harvard University Press, 1991).
2. Ibid.; U.N. Environment Programme (UNEP), *Environmental Effects of Ozone Depletion: 1994 Assessment* (Nairobi: 1994).
3. Edward A. Parson and Owen Greene, "The Complex Chemistry of the International Ozone Agreements," *Environment*, March 1995; Hilary F. French, "Learning from the Ozone Experience," in Lester R. Brown et al., *State of the World 1997* (New York: W.W. Norton & Company, 1997); current control requirements from UNEP, Industry and Environment Office, "OzonAction Information Clearinghouse Database," < http://www.epa.gov/docs/ozone/intpol/oaic.html >, viewed 7 November 1996.
4. UNEP, op. cit. note 3.
5. Douglas G. Cogan, *Stones in a Glass House: CFCs and Ozone Depletion* (Washington, DC: Investor Responsibility Research Center, 1988).
6. Decline of 76 percent based on DuPont data cited in Anjali Acharya, "CFC Production Drop Continues," in Lester R. Brown, Christopher Flavin, and Hal Kane, *Vital Signs 1996* (New York: W.W. Norton & Company, 1996).
7. UNEP, "The Reporting of Data by the Parties to the Montreal Protocol on Substances That Deplete the Ozone Layer" (Nairobi: 12 September 1996).
8. Elizabeth Cook, *Marking a Milestone in Ozone Protection: Learning from the CFC Phase-Out* (Washington, DC: World Resources Institute, January 1996).
9. Duncan Brack, *International Trade and the Montreal Protocol* (London: Earthscan, for the Royal Institute for International Affairs, 1996).
10. Owen Greene, "The Montreal Protocol: Implementation and Development in 1995," in J. Poole and R. Guthrie, eds., *Verification 1996: Arms Control, Environment, and Peacekeeping* (Boulder, CO: Westview Press, 1996).
11. Brack, op. cit. note 9.
12. Ibid.
13. UNEP, op. cit. note 3.
14. "Parties to Montreal Protocol Agree to Phase Out Methyl Bromide by 2010," *International Environment Reporter*, 13 December 1995.
15. UNEP, op. cit. note 3.
16. Ibid.
17. Multilateral Fund Secretariat, "Country Programme Summary Sheets" (Montreal: October 1996).
18. UNEP, op. cit. note 7.
19. Ibid.
20. Ibid.
21. World Meteorological Organization (WMO), *Scientific Assessment of Ozone Depletion: 1994* (Geneva: 1995).
22. Ibid.
23. Ibid.; D.J. Hofmann, "Recovery of Antarctica Ozone Hole" (letter to editor), *Nature*, 21 November 1996.
24. WMO, "Stronger Ozone Decline Continues," press release (Geneva: 12 March 1996); Elizabeth Olson, *WMO Antarctic Ozone Bulletin*, November 1996, < http://www.wmo.ch/web/press/press.html >, posted 11 December 1996; WMO, op. cit. note 21; J.R. Herman et al., "UV-B Increases (1979–1992) from Decreases in Total Ozone," *Geophysical Research Letters*, 1 August 1996.

HARMFUL SUBSIDIES WIDESPREAD (pages 104–05)

1. Figure of $500 billion is a Worldwatch estimate, based on sources for Table 1: mining subsidies from U.S. Congress, Committee on Natural Resources, Subcommittee on Oversight and Investigations, *Taking from the Taxpayer: Public Subsidies for Natural Resource Development*, Majority Staff Report (Washington, DC: 1994), from Juri Peepre, Canadian Parks and Wilderness Society, White Horse, Yukon, discussion with author, 24 June 1996, and from David N. Smith and Louis T. Wells, Jr., *Negotiating Third-World Mineral Agreements: Promises as Prologue* (Cambridge, MA: Ballinger, 1975); logging subsidies from Theodore Panayotou and Peter S. Ashton, *Not By Timber Alone: Economics and Ecology for Sustaining Tropical Forests* (Washington, DC: Island Press, 1992), from Randal O'Toole, "Timber Sale Subsidies, But Who Gets Them?" *Different Drummer* (Thoreau Institute, Oak Grove, OR), spring 1995, and from Andrew K. Dragun, "The

Subsidization of Logging in Victoria," unpublished paper (Melbourne: LaTrobe University, 1995); fishing subsidies based on Rory McLeod, *Market Access Issues for the New Zealand Seafood Trade* (Wellington: New Zealand Fishing Industry Board, 1996), and on U.N. Food and Agriculture Organization (FAO), *Marine Fisheries and the Law of the Sea: A Decade of Change*, FAO Fisheries Circular No. 853 (Rome: 1993); irrigation subsidies from Gregory K. Ingraham and Marianne Fay, "Valuing Infrastructure Stocks and Gains from Improved Performance," background paper for *World Development Report 1994* (Washington, DC: World Bank, 1994), and from U.S. Congress, op. cit. this note; fertilizer subsidies from Sanjeev Gupta, Kenneth Miranda, and Ian Parry, *Public Expenditure Policy and the Environment: A Review and Synthesis*, IMF Working Paper (Washington, DC: International Monetary Fund (IMF), 1993); pesticide subsidies from Jumanah Farah, *Pesticide Policies in Developing Countries: Do They Encourage Excessive Use?* World Bank Discussion Paper 238 (Washington, DC: World Bank, 1994); crop production subsidies from Organisation for Economic Co-operation and Development (OECD), *Agricultural Policies, Markets and Trade in OECD Countries* (Paris: 1996); livestock production subsidies from Alan B. Durning and Holly B. Brough, *Taking Stock: Animal Farming and the Environment*, Worldwatch Paper 103 (Washington, DC: Worldwatch Institute, July 1991); energy use subsidies from Bjorn Larsen and Anwar Shah, "Global Climate Change, Energy Subsidies and National Carbon Taxes," in Lans Bovenberg and Sijbren Cnossen, eds., *Public Economics and the Environment in an Imperfect World* (Boston: Kluwer Academic Press, 1995), from Giancarlo Tosato, *Environmental Implications of Support to the Electric Sector in Italy: A Case Study*, Preliminary Draft Report to OECD Environment Directorate (Paris: OECD, 1995), cited in Laurie Michaelis, "The Environmental Implications of Energy and Transport Subsidies," in OECD, *Subsidies and Environment: Exploring the Linkages* (Paris: 1996), from Douglas Koplow, *Federal Energy Subsidies: Energy, Environmental, and Fiscal Impacts* (Washington, DC: Alliance to Save Energy, 1993), and from U.S. Department of Energy,

Energy Information Administration, *Federal Energy Subsidies: Direct and Indirect Interventions in Energy Markets* (Washington, DC: U.S. Government Printing Office, 1992); driving subsidy is a Worldwatch estimate, based on difference between road spending and road user fees from U.S. Department of Transportation (DOT), Federal Highway Administration (FHA), *Highway Statistics 1994* (Washington, DC: 1995), < http://www.bts. gov/fhwa/yellowbook/section4/hf10.xls >, viewed 26 December 1996, on the value of the tax break for employer-provided free parking from *Transportation Sector Subsidies: U.S. Case Study*, prepared for U.S. Environmental Protection Agency (Lexington, MA: DRI/McGraw-Hill, 1994), and on the cost of road-related services from James J. MacKenzie, Roger C. Dower, and Donald D.T. Chen, *The Going Rate: What It Really Costs to Drive* (Washington, DC: World Resources Institute (WRI), 1992). For further details, see David Malin Roodman, *Paying the Piper: Subsidies, Politics, and the Environment*, Worldwatch Paper 133 (Washington, DC: Worldwatch Institute, December 1996).

2. Robert Repetto, *The Forest for the Trees? Government Policies and the Misuse of Forest Resources* (Washington, DC: WRI, 1988).

3. Ronald P. Steenblik and Panos Coroyannikis, "Reform of Coal Policies in Western and Central Europe: Implications for the Environment," *Energy Policy* 23, no. 6 (1995).

4. Effects in the United States from Jonathan Tolman, *Federal Agricultural Policy: A Harvest of Environmental Abuse* (Washington, DC: Competitive Enterprise Institute, 1996); effects in Western Europe from C. Ford Runge, "The Environmental Impacts of Agricultural and Forest Subsidies," in OECD, *Subsidies and Environment*, op. cit. note 1.

5. See, for example, Charles F. Wilkinson, *Crossing the Next Meridian: Land, Water, and the Future of the West* (Washington, DC: Island Press, 1992).

6. Peepre, op. cit. note 1; U.S. Congress, op. cit. note 1.

7. Thomas J. Hilliard, *Golden Patents, Empty Pockets: A 19th Century Law Gives Miners Billions, the Public Pennies* (Washington, DC: Mineral Policy Center, 1994).

8. Dragun, op. cit. note 1. Figure cited excludes value of environmental and recreational losses.

9. O'Toole, op. cit. note 1.

10. Charles Victor Barber, Nels C. Johnson, and Emmy Hafild, *Breaking the Logjam: Obstacles to Forest Policy Reform in Indonesia and the United States* (Washington, DC: WRI, 1994); aid figure from World Bank, *World Development Report 1993* (New York: Oxford University Press, 1993).

11. Barber, Johnson, and Hafild, op. cit. note 10.

12. Jeffrey D. Sachs and Andrew M. Warner, *Natural Resource Abundance and Economic Growth*, Development Discussion Paper No. 517a (Cambridge, MA: Harvard Institute for International Development, October 1995).

13. Subsidy figure from OECD, International Energy Agency, *Energy Policies of IEA Countries* (Paris: 1996), converted using a 1995 exchange rate; employment figures from idem, *Coal Information* (Paris: various years).

14. FAO, op. cit. note 1; Peter Weber, *Net Loss: Fish, Jobs, and the Marine Environment*, Worldwatch Paper 120 (Washington, DC: Worldwatch Institute, July 1994).

15. Subsidy program reach from Christine Kerr and Leslie Citroen, "Household Expenditures on Infrastructure Services," background paper for *World Development Report 1994* (Washington, DC: World Bank, undated); subsidy distribution from Einar Hope and Balbir Singh, *Energy Price Increases in Developing Countries: Case Studies of Colombia, Ghana, Indonesia, Malaysia, Turkey, and Zimbabwe*, Policy Research Working Paper 1442 (Washington, DC: World Bank, 1995).

16. Figure of $111 billion is explained in note 1.

17. Price rise is a Worldwatch estimate based on a motor fuel usage rate from DOT, FHA, *Highway Statistics 1994* (Washington, DC: 1995), < http://www.bts.gov/fhwa/yellowbook/section1/mf21.xls >, viewed 26 December 1996; effects from MacKenzie, Dower, and Chen, op. cit. note 1.

18. Figure of $7.5 trillion is a Worldwatch estimate, based on gross domestic product (GDP) figures from World Bank, *World Data 1994: World Bank Indicators on CD-ROM*, electronic database (Washington, DC: 1994), on western industrial-country figure from OECD, *Revenue Statistics of OECD Member Countries 1960–1994* (Paris: 1995), on taxes as a share of GDP for former Eastern bloc countries from IMF, *World Economic Outlook, October 1994* (Washington, DC: 1994), and on central government revenues (including nontax revenues) as a share of GDP for developing countries from idem, *Government Finance Statistics Yearbook 1994* (Washington, DC: 1994).

SUSTAINABLE DEVELOPMENT AID THREATENED (pages 108–09)

1. Michael Grubb et al., *The Earth Summit Agreements: A Guide and Assessment* (London: Earthscan Publications Ltd., 1993).

2. Organisation for Economic Co-operation and Development (OECD), *Development Co-operation*, 1996 Report of the Development Assistance Committee (Paris: 1997).

3. Ibid.

4. Ibid.

5. See chapter 33 in United Nations, *Agenda 21: The United Nations Programme of Action from Rio* (New York: U.N. Publications, 1992).

6. The United Nations Framework Convention on Climate Change and the Convention on Biological Diversity are included in Lakshman D. Guruswamy, Sir Geoffrey W. R. Palmer, and Burns H. Weston, *International Environmental Law and World Order*, Supplement of Basic Documents (St. Paul, MN: West Publishing Co., 1994).

7. United Nations, op. cit. note 5.

8. OECD, op. cit. note 2.

9. United Nations, op. cit. note 5.

10. Patti L. Petesch, *North-South Environmental Strategies, Costs, and Bargains*, Policy Essay No. 5 (Washington, DC: Overseas Development Council, 1992).

11. OECD, op. cit. note 2.

12. Ibid.

13. Congressional Research Service (CRS), "Foreign Operations Appropriations for FY997: Funding and Policy Issues," U.S. Library of Congress, Washington, DC, updated 1 October 1996.

14. Ibid.

15. OECD, op. cit. note 2.

16. United Nations, op. cit. note 5.

17. Grubb et al., op. cit. note 1; size of U.S. contribution from World Bank, *World Bank: Annual Report 1996* (Washington, DC: 1996); CRS, op. cit. note 13.

18. Global Environment Facility (GEF), brochure, Washington, DC, December 1991.

19. "Agreement Reached on Funding GEF; Program to Receive More than $2 Billion," *International Environment Reporter*, 23 March 1994; GEF, "Instrument for the Establishment of the Restructured Global Environment Facility," Report of the GEF Participants Meeting, Geneva, Switzerland, 14–16 March 1994.

20. Seth Dunn, "The Berlin Climate Change Summit: Implications for International

Environmental Law," *International Environment Reporter*, 31 May 1995; "Global Environment Facility To Continue as 'Interim' Financing Source for Projects," *International Environment Reporter*, 14 December 1994.

21. Figures on U.S. budget requests and appropriations for the GEF from CRS, op. cit. note 13; Frederik van Bolhuis, Senior Environmental Economist, GEF, Washington, DC, discussion with author, 8 January 1997.

22. Van Bolhuis, op. cit. note 21.

23. Anne Bohon, GEF Secretariat, Washington, DC, discussion with author, 23 August 1996.

24. *United Nations Environment Programme (UNEP), Sources of Funds, 1992–1996*, provided by Jim Sniffen, U.N. Environment Programme, New York, discussion with author, 19 February 1997; *Annual Report of the Administrator for 1995 and Program Level Assistance Statistics*, Statistical Annex, DP/1996/18/Add.4 (29 April 1996), provided by Rameesh Gampat, U.N. Development Programme, New York, discussion with author, 18 February 1997.

25. Sniffen, op. cit. note 24; the figures do not include contributions to the Montreal Protocol trust fund.

26. Gampat, op. cit. note 24.

27. Shanti R. Conly, *The United Nations and Population Assistance* (Washington, DC: Population Action International, 1996); UNFPA, *Report 1995* (New York: 1995).

28. Conly, op. cit. note 27; Population Action International, 1995 *Annual Report* (Washington, DC: 1996).

29. Conly, op. cit. note 27.

FOOD AID FALLS SHARPLY
(pages 110–11)

1. U.N. Food and Agriculture Organization (FAO), "FAOSTAT DATA," < http://www.fao.org >, Rome, viewed 23 September 1996; 90 percent from Shalha Shipouri and Margaret Missiaen, "Shortfalls in International Food Aid Expected," *FoodReview*, September–December 1995.

2. Shipouri and Missiaen, op. cit. note 1.

3. Ibid.

4. Jay Sjerven, "Slashes in Food Aid Raise Concerns About U.S. Commitment," *Milling and Baking News* 75, no. 3 (1996). Concessional sales are those at least 25 percent below the commercial price, according to David Phiri,

FAO, e-mail message to author, 22 November 1996.

5. FAO, "Food Security and Food Assistance," in *Technical Background Documents 12-15: Vol. 3* (Rome: 1996).

6. Ibid.

7. Shapouri and Missiaen, op. cit. note 1.

8. FAO, op. cit. note 1.

9. Sjerven, op. cit. note 4.

10. Shapouri and Missiaen, op. cit. note 1.

11. Calculated from data in FAO, op. cit. note 1.

12. Ibid.

13. Shapouri and Missiaen, op. cit. note 1.

14. Ibid.

15. FAO, op. cit. note 5.

16. FAO, op. cit. note 1; aid share of imported food from FAO, op. cit. note 5.

17. Shipouri and Missiaen, op. cit. note 1.

18. Catherine Gwin, "Reforming Aid," *Ceres*, January/February 1996.

19. FAO, op. cit. note 5.

20. Steven Greenhouse, "U.N. Aid Officials Plead for Food Relief," *New York Times*, 30 April 1995.

21. Lester R. Brown, "Facing the Prospect of Food Scarcity," in Lester R. Brown et al., *State of the World 1997* (New York: W.W. Norton & Company, 1997), and Lester R. Brown, "Grain Stocks Drop to All-Time Low," ·in Lester R. Brown, Christopher Flavin, and Hal Kane, *Vital Signs 1996* (New York: W.W. Norton & Company, 1996).

22. Worldwatch calculation based on food aid totals from FAO and grain production totals from the U.S. Department of Agriculture (USDA). Even in 1992, the peak year of giving, grain food aid constituted only 0.4 percent of global grain production.

23. Thomas Sheehy, "Who Has Aid Really Helped?" *Ceres*, January/February 1996.

24. David Ransom, "The Poverty of Aid," *The New Internationalist*, November 1996.

25. Gwin, op. cit. note 18.

26. Sjerven, op. cit. note 4.

27. Shapouri and Missiaen, op. cit. note 1.

28. Ibid.

29. USDA, *Food Aid Needs and Availabilities: Projections to 2005* (Washington, DC: Economic Research Service, October 1995).

30. Ibid.

31. Ibid.

R&D SPENDING LEVELS OFF
(pages 112–13)

1. National Science Board (NSB), *Science and Engineering Indicators 1996* (Washington, DC: U.S. Government Printing Office, 1996).
2. OECD members' national currencies were translated into dollars by using purchasing-power parities (PPP) rather than market exchange rates; calculated on basis of Organisation for Economic Co-operation and Development (OECD), *Main Science and Technology Indicators, 1996/1* (Paris: 1996).
3. NSB, op. cit. note 1.
4. "The Shrinking Science Budget," *New York Times*, 23 January 1997.
5. Andrew Pollack, "Japan Is Planning Vast Increase in Science Research Budget," *New York Times*, 2 July 1996.
6. OECD, op. cit. note 2.
7. NSB, op. cit. note 1.
8. Ibid.; U.N. Educational, Scientific and Cultural Organization, *UNESCO Statistical Yearbook 1995* (Lanham, MD: Bernan Press, 1995).
9. Calculated from Pacific Economic Cooperation Council, Science and Technology Task Force, "APEC/PECC Pacific Science and Technology Profile," Fourth Issue 1995. National currencies were translated into dollars by using purchasing-power parities.
10. Ibid.; NSB, op. cit. note 1. Rubles were translated into dollars by using purchasing power parities.
11. OECD, op. cit. note 2.
12. Ibid.
13. Ibid.
14. Bonn International Center for Conversion (BICC), *Conversion Survey 1996. Global Disarmament, Demilitarization and Demobilization* (New York: Oxford University Press, 1996).
15. Ibid.; OECD, op. cit. note 2.
16. NSB, op. cit. note 1.
17. Ibid.
18. International Energy Agency (IEA), *Energy Policies of IEA Countries. 1996 Review* (Paris: 1996); IEA, *IEA Energy Technology R&D Statistics, 1974–1995* (Paris: forthcoming, 1997), as provided in electronic version by Karen Treanton, IEA, Paris, e-mail message to author, 22 January 1997.
19. IEA, *1996 Review*, op. cit. note 18.
20. IEA, *R&D Statistics*, op. cit. note 18.
21. Calculated on basis of IEA, *1996 Review*, op. cit. note 18.
22. IEA, *R&D Statistics*, op. cit. note 18.
23. Ibid.
24. U.N. Food and Agriculture Organization (FAO), "Food Security and Agricultural Research," pamphlet issued for the World Food Summit (Rome: 13–17 November 1996).
25. Ibid.
26. FAO, *Food for All*, published on the occasion of the World Food Summit (Rome: 13–17 November 1996).
27. NSB, op. cit. note 1.
28. Ibid.
29. Ibid.
30. Ibid.
31. Ibid.
32. Ibid.

AGRICULTURE GROWS IN CITIES
(pages 114–15)

1. Table 1 based on the following sources: Kampala from Daniel G. Maxwell, "Alternative Food Security Strategy: A Household Analysis of Urban Agriculture in Kampala," *World Development* 23, no. 10 (1995); Havana from John McKenzie, "Revolutionary Food," *Habitat Australia*, June 1996, and from Adam Tiller, Permaculture Global Assistance Network, Melbourne, Australia, e-mail message to author, 4 December 1996; East Calcutta and Singapore from United Nations Development Programme (UNDP), *Urban Agriculture: Food, Jobs and Sustainable Cities* (New York: 1996); Hartford from Rachel Nugent, "The Sustainability of Urban Agriculture: A Case Study in Hartford, Connecticut," Environmental Science and Engineering Fellows Program (Washington, DC: American Association for the Advancement of Science, 1996); Berlin from Patricia Hynes, *A Patch of Eden* (White River Junction, VT: Chelsea Green Publishing Company, 1996).
2. UNDP, op. cit. note 1.
3. World Resources Institute (WRI) et al., *World Resources 1996–97* (New York: Oxford University Press, 1996); United Nations, *World Urbanization Prospects: The 1994 Revision* (New York: 1995).
4. WRI et al., op. cit. note 3; A. Cecilia Snyder, *Hunger Facts*, Bread for the World Background Paper No. 124 (Silver Spring, MD: February 1994).

5. Snyder, op. cit. note 4.
6. Luc J.A. Mougeot, "African City Farming from a World Perspective," in International Development Research Centre (IDRC), *Cities Feeding People: An Examination of Urban Agriculture in East Africa* (Ottawa, ON, Canada: 1994).
7. Camillus J. Sawio, *Urban Agriculture Research in East and Central Africa: Records, Capacities and Opportunities*, Cities Feeding People Series, Report 10 (Ottawa, ON, Canada: IDRC, 1993); "Breaking New Ground in Dar es Salaam," *Urban Perspectives*, November 1993.
8. UNDP, op. cit. note 1.
9. McKenzie, op. cit. note 1; Tiller, op. cit. note 1.
10. "Breaking New Ground," op. cit. note 7.
11. UNDP, op. cit. note 1.
12. Pablo Gutman, "Feeding the City: Potential and Limits of Self-Reliance," *Development: Seeds of Change* 4 (1986).
13. WRI et al., op. cit. note 3.
14. Ibid.
15. Kristin Helmore and Annu Ratta, "The Surprising Yields of Urban Agriculture," *Choices* (UNDP), April 1995.
16. Ibid.
17. UNDP, op. cit. note 1; WRI et al., op. cit. note 3.
18. "Wastewater Use in Aquaculture: Research in Peru," *NAGA, The ICLARM Quarterly*, October 1995.
19. WRI et al., op cit. note 3; UNDP, op. cit. note 1.
20. WRI et al., op cit. note 3.
21. Jac Smit, The Urban Agriculture Network, Washington, DC, e-mail message to author, 14 September 1996.
22. Rachel Nugent, "Urban Agriculture: An Oxymoron?" in U.N. Food and Agriculture Organization, *State of Food and Agriculture* (Rome: 1996).

GAP IN INCOME DISTRIBUTION WIDENING (pages 116–17)

1. U.N. Development Programme (UNDP), *Human Development Report 1996* (New York: Oxford University Press, 1996).
2. Ibid.
3. Ibid.
4. Robin Wright, "The Gap Between Haves and Have-Nots Is Growing," *21st Century Earth* (San Diego, CA: Greenhaven Press, 1996).

5. Khozem Merchant, "World `Heads for Grotesque Inequalities,'" *Financial Times*, 16 July 1996.
6. UNDP, op. cit. note 1.
7. Diana Jean Schemo, "Brazil's Chief Acts to Take Land to Give To the Poor," *New York Times*, 13 November 1995.
8. UNDP, op. cit. note 1.
9. Ibid.
10. Ibid.
11. Ibid.
12. Ibid.
13. Ibid.
14. Ibid.
15. Ibid.
16. Ibid.
17. Ibid.
18. Ibid.
19. Steven A. Holmes, "Income Disparity Between Poorest and Richest Rises," *New York Times*, 20 June 1996.
20. Ibid.
21. Ibid.
22. Ibid.
23. UNDP, op. cit. note 1.
24. The studies are by Alberto Alesina at Harvard University and Roberto Perotti at Columbia University, cited in Keith Bradsher, "More on the Wealth of Nations," *New York Times*, 20 August 1995.
25. Bradsher, op. cit. note 24.
26. Ibid.

ELECTRIC CARS HIT THE ROAD (pages 118–19)

1. Worldwatch estimate based on International Energy Agency (IEA), "Implementing Agreement for Electric Vehicle Technologies and Programmes" (Paris: Organisation for Economic Co-operation and Development, July 1996), on European Electric Road Vehicle Association (AVERE), "Electric Vehicles On-road in Europe at the End of 1995" (Brussels: 26 November 1996), and on Electric Vehicle Association of the Americas (EVAA), "Electric Vehicle Population in the United States" (San Francisco: March 1995).
2. AVERE, op. cit. note 1.
3. IEA, op. cit. note 1.
4. Ibid.
5. AVERE, op. cit. note 1.

6. IEA, op. cit. note 1.
7. EVAA, op. cit. note 1.
8. Natural Resources Defense Council (NRDC), *Green Auto Racing: National Efforts and International Cooperation to Promote Advanced Cars and Fuels* (Washington, DC: July 1996).
9. Matthew L. Wald, "Electric Cars in California Are Set Back," *New York Times*, 30 March 1996.
10. Ibid.
11. B. Drummond Ayres, Jr., "No Engine, No Gasoline, But a Car Nonetheless," *New York Times*, 6 December 1996; Chris Kraul, "Honda, Ford Unveil Plans for EV Models," *Los Angeles Times*, 3 January 1997.
12. CALSTART, *Electric Vehicles: An Industry Prospectus* (Burbank, CA: October 1996).
13. NRDC, op. cit. note 8.
14. Ibid.
15. CALSTART, op. cit. note 12.
16. Li Yan, "Electric Vehicles Can Power Industry," *China Daily*, 9 December 1996; "Peugeot Electric Cars in China," *Wall Street Journal*, 31 December 1996.
17. Richard de Neufville et al., "The Electric Car Unplugged," *Technology Review*, January 1996; "Turning Up the Heat," *Consumer Reports*, September 1996.
18. Amal Kumar Naj, "You Can Buy Yourself an Electric Car, But It Isn't Going to Take You Very Far," *Wall Street Journal*, 15 May 1996.
19. Daniel Sperling, "The Case for Electric Vehicles," *Scientific American*, November 1996.
20. Global Development and Environment Institute, *Near-Term Electric Vehicle Costs* (Medford, MA: Tufts University, May 1994).
21. William J. Cook, "Look, Mom, No Gas," *U.S. News and World Report*, 30 September 1996.
22. Taylor Moore, "The Road Ahead," *EPRI Journal*, March/April 1996.
23. Lawrence Fisher, "G.M., in a First, Will Sell a Car Designed for Electric Power This Fall," *New York Times*, 5 January 1996.
24. Daniel Sperling, *Future Drive: Electric Vehicles and Sustainable Transportation* (Washington, DC: Island Press, 1995).
25. CALSTART, op. cit. note 12.
26. Ibid.
27. Ibid.
28. Ibid.
29. Ibid.; NRDC, op. cit. note 8.
30. CALSTART, op. cit. note 12; NRDC, op. cit. note 8.

ARMS PRODUCTION FALLS
(pages 120–21)

1. Calculated from International Institute for Strategic Studies (IISS), *The Military Balance 1996/97* (London: Oxford University Press, October 1996), and from IISS, *The Military Balance 1993–1994* (London: Brassey's, October 1993).
2. Calculated from IISS, *The Military Balance 1996/97*, op. cit. note 1, and from IISS, *The Military Balance 1993–1994*, op. cit. note 1.
3. Calculated from IISS, *The Military Balance 1996/97*, op. cit. note 1, and from IISS, *The Military Balance 1993–1994*, op. cit. note 1. In 1993–95, the signatories of the Conventional Forces in Europe Treaty were expected to cut their arsenals to negotiated lower ceilings.
4. Michael Renner, *Cost of Disarmament: An Overview of the Economic Costs of the Dismantlement of Weapons and the Disposal of Military Surplus*, Brief 6 (Bonn, Germany: Bonn International Center for Conversion (BICC), March 1996).
5. BICC, *Conversion Survey 1996. Global Disarmament, Demilitarization and Demobilization* (New York: Oxford University Press, 1996).
6. Ibid.
7. Ibid.
8. Ibid.
9. Data for 1994 from Elisabeth Sköns and Bates Gill, "Arms Production," in Stockholm International Peace Research Institute (SIPRI), *SIPRI Yearbook 1996: Armaments, Disarmament and International Security* (New York: Oxford University Press, 1996); 1990 data from Paolo Miggiano, Elisabeth Sköns, and Herbert Wulf, "Arms Production," in SIPRI, *SIPRI Yearbook 1992: World Armaments and Disarmament* (New York: Oxford University Press, 1992). These figures are not fully comparable because the precise composition of the top 100 changes somewhat from year to year, but they serve as an adequate overall trend indicator.
10. Sköns and Gill, op. cit. note 9; BICC, op. cit. note 5.
11. Worldwatch estimate based on Sköns and Gill, op. cit. note 9, and on BICC, op. cit. note 5.
12. BICC, op. cit. note 5.
13. Ibid.
14. Ibid.
15. Ibid.
16. Ibid.

17. Christopher Louise, *The Social Impacts of Light Weapons Availability and Proliferation*, Discussion Paper No. 59 (Geneva: U.N Research Institute for Social Development, March 1995).
18. U.S. Arms Control and Disarmament Agency (ACDA), *World Military Expenditures and Arms Transfers 1996* (Washington, DC: U.S. Government Printing Office, forthcoming).
19. ACDA, *World Military Expenditures and Arms Transfers 1995* (Washington, DC: U.S. Government Printing Office, April 1996); ACDA, op. cit. note 18.
20. ACDA, op. cit. note 19.
21. Ibid.
22. Ibid.
23. Ibid.
24. BICC, op. cit. note 5.
25. SIPRI estimate from ibid.
26. Ibid.
27. Ibid.
28. Ibid.

GLOBAL POPULATION GROWING OLDER (pages 124–25)

1. U.S. Bureau of the Census, *World Population Prospects: 1996* (Washington, DC: U.S. Government Printing Office, July 1996); United Nations, *World Population Prospects: The 1996 Revision*, Annex I (New York: forthcoming).
2. United Nations, op. cit. note 1.
3. United Nations, *World Population Prospects: The 1996 Revision,* Annex II & III (New York: forthcoming).
4. Ibid.; U.S. Bureau of the Census, *An Aging World II* (Washington, DC: U.S. Government Printing Office, 1992).
5. United Nations, op. cit. note 3; Bureau of the Census, op. cit. note 4.
6. United Nations, op. cit. note 3.
7. Ibid.
8. Ibid.
9. He Changmei, "Situation of Population Aging in China and the Strategy," *China Population Today*, August 1996; United Nations, op. cit. note 3.
10. Atoh Makoto, "Fewer Children, More Seniors," *Japan Echo*, Special Issue 1996; Ogawa Naohiro, "When The Baby Boomers Grow Old," *Japan Echo*, Special Issue 1996.
11. Hong Guodong, "Aging of Populations: Trends and Countermeasures," *Beijing Review*, 12–18

February 1996; He, op. cit. note 9; Bureau of the Census, op. cit. note 1; United Nations, op. cit. note 1.
12. "New Populations of Old Add to Poor Nations' Burdens," *Science*, 5 July 1996; United Nations, op. cit. note 3.
13. United Nations, op. cit. note 3.
14. Bureau of Census, op. cit. note 4.
15. Nicholas D. Kristof, "Aging World, New Wrinkles," *New York Times*, 22 September 1996.
16. Peter Hicks, "The Impact of Aging on Public Policy," *OECD Observer*, December 1996/January 1997.
17. Kristof, op. cit. note 15.
18. Jessica Mathews, "Retirement Crisis," *Washington Post*, 1 January 1995; United Nations, op. cit. note 1.
19. Urie Bronfenbrenner et al., *The State of Americans* (New York: The Free Press, 1996); Bureau of the Census, op. cit. note 4.
20. Steven Mufson, "China's Raging Silver Wave," *Washington Post*, 7 August 1995.
21. Hong, op. cit. note 11; Wu Naitao, "Community Services for Senior Citizens," *Beijing Review*, 12–18 February 1996.
22. "New Populations of Old," op. cit. note 12.
23. Bureau of the Census, op. cit. note 4; Mark Suzman, "Illness Clouds Longer Life Expectancy," *Financial Times*, 30 January 1997.
24. Mathews, op. cit. note 18.
25. "New Populations of Old," op. cit. note 12.
26. United Nations, op. cit. note 1; Bureau of the Census, op. cit. note 4.
27. United Nations, op. cit. note 1.
28. Margot Jefferys, "A New Way of Seeing Old Age is Needed," *World Health*, September–October 1996.
29. Stephenie Flanders, "Conflict Between Age Group Looms," *Financial Times*, 27 September 1996.
30. Bureau of the Census, op. cit. note 4.
31. Ibid.
32. Marvin Kaiser, "Productive Roles of Older Individuals in Developing Countries," *Generations*, Winter 1993.

NONCOMMUNICABLE DISEASES RISING (pages 126–27)

1. Christopher J.L. Murray and Alan D. Lopez, "Global Patterns of Cause of Death and Burden of Disease in 1990, with Projections to 2020," in

World Health Organization (WHO), Report of the Ad Hoc Committee on Health Research Relating to Future Intervention Options, *Investing in Health Research and Development* (Geneva: 1996).

2. Ibid.

3. Harvard School of Public Health, World Bank, and WHO, "Summary: Global Burden of Disease and Injury Series," in Christopher J.L. Murray and Alan D. Lopez, eds., *The Global Burden of Disease* (Cambridge, MA: Harvard University Press, 1996).

4. Ibid.

5. WHO, Tobacco or Health Program, "The Tobacco Epidemic: A Global Public Health Emergency," in *Tobacco Alert: Special Issue*, World No-Smoke Day 1996.

6. "Fatal Fags," *Down to Earth*, 15 January 1995.

7. Murray and Lopez, op. cit. note 1.

8. Ibid.

9. Paolo Boffetta and D. Maxwell Parkin, "Cancer in Developing Countries," *CA: A Cancer Journal for Clinicians*, March/April 1994.

10. WHO, op. cit. note 1.

11. Shanghai County from Kenneth Stanley, "Control of Tobacco Production and Use," Appendix A, in Dean T. Jamison et al., eds., *Disease Control Priorities in Developing Countries* (Washington, DC: World Bank, 1993).

12. Ibid.

13. Ibid.

14. John Briscoe, *Brazil: The New Challenge of Adult Health*, A World Bank Country Study (Washington, DC: August 1990).

15. Ibid.

16. Ibid.

17. Murray and Lopez, op. cit. note 1.

18. Ibid.

19. National Research Council, *Diet and Health: Implications for Reducing Chronic Disease Risk* (Washington, DC: National Academy Press, 1989).

20. Vladimir Kebza and Rudolf Poledne, "Smoking Cessation Models Bring Results in the Czech Republic," *Central European Health and Environment Monitor*, Fall/Winter 1995.

21. Ibid.

22. World Bank, *World Development Report 1993* (New York: Oxford University Press, 1993).

MATERNAL MORTALITY REMAINS HIGH (pages 128–29)

1. The new estimate is a recalculation for 1990 and reveals that an additional 80,000 maternal deaths occur each year than previous estimates suggested; World Health Organization (WHO), *Revised 1990 Estimates of Maternal Mortality: A New Approach by WHO and UNICEF* (Geneva: 1996).

2. WHO, *Maternal Mortality: A Global Factbook* (Geneva: 1991).

3. Ibid.

4. U.N. Population Division, *Review and Appraisal of the World Population Plan of Action: 1994 Report* (New York: United Nations, 1995); James McCarthy and Deborah Maine, "A Framework for Analyzing the Determinants of Maternal Mortality," *Studies in Family Planning*, January/February 1992.

5. WHO, op. cit. note 1.

6. Ibid.

7. Ibid.

8. McCarthy and Maine, op. cit. note 4; U.N. Population Division, op. cit. note 4.

9. U.N. Population Fund (UNFPA), *1996 Population Issues Briefing Kit* (New York: 1996).

10. WHO, op. cit. note 1.

11. WHO, op. cit. note 2; Erica Royston and Sue Armstrong, eds., *Preventing Maternal Deaths* (Geneva: WHO, 1989); U.N. Population Division, op. cit. note 4.

12. WHO, op. cit. note 2; Royston and Armstrong, op. cit. note 11; U.N. Population Division, op. cit. note 4.

13. U.N. Population Division, op. cit. note 4.

14. Peter Adamson, "A Failure of Imagination," in UNICEF, *The Progress of Nations 1996* (Geneva: 1996)

15. Julia A. Walsh et al., "Maternal and Perinatal Health," in Dean T. Jamison et al., eds., *Disease Control Priorities in Developing Countries* (New York: Oxford University Press, 1993); Ann Starrs, *Preventing the Tragedy of Maternal Deaths: A Report on the International Safe Motherhood Conference* (Nairobi: WHO, 1987).

16. Walsh et al., op. cit. note 15; Starrs, op. cit. note 15.

17. Starrs, op. cit. note 15.

18. Adriana Gómez, "Campaign to Prevent Maternal Mortality and Morbidity: Evaluating the Past Eight Years," *Women's Health Journal*, January 1996.

19. McCarthy and Maine, op. cit. note 4.
20. World Bank estimate cited in Royston and Armstrong, op. cit. note 11.
21. United Nations, *Programme of Action of the U.N. International Conference on Population and Development* (New York: 1994).
22. J. Ties Boerma, "Levels of Maternal Mortality in Developing Countries," *Studies in Family Planning*, July/August 1987; U.N. Population Division, op. cit. note 4; *Safe Motherhood Newsletter*, Winter 1992/93.
23. McCarthy and Maine, op. cit. note 4; Adamson, op. cit. note 14; Starrs, op. cit. note 15.
24. U.N. Population Division, op. cit. note 4.

HALF OF LANGUAGES BECOMING EXTINCT (pages 130–31)

1. Nancy Lord, "Native Tongues: The Languages That Once Mapped the American Landscape Have Almost Vanished," *Sierra*, November/December 1996.
2. Michael Krauss, "The World's Languages in Crisis," *Language* 68, no. 1, 1992.
3. Ibid.
4. Ibid.
5. Jared Diamond, "Speaking With A Single Tongue," *Discover*, February 1993.
6. Ibid.
7. Ibid.
8. Michael Krauss, "A Loss For Words," *Simply Living*, August 1992.
9. Ibid.
10. Ibid.
11. Krauss, op. cit. note 2.
12. Diamond, op. cit. note 5.
13. Ibid.
14. Catherine Gysin, "Language: The Lost Horizon," *Utne Reader*, May/June 1993.
15. Krauss, op. cit. note 2.
16. Krauss, op. cit. note 8.
17. Ibid.
18. Lord, op. cit. note 1.
19. Ibid.
20. Ibid.
21. Burkhard Bilger, "Keeping Our Words," *The Sciences*, September/October 1994.
22. Krauss, op. cit. note 2.
23. Ibid.
24. Diamond, op. cit. note 5; Krauss, op. cit. note 2.
25. Diamond, op. cit. note 5.
26. Ibid.
27. Ibid.
28. Krauss, op. cit. note 2.
29. Diamond, op. cit. note 5.
30. Ibid.
31. Ibid.
32. Sally Weeks, "Why English? Because It Works," *The WorldPaper*, July 1996.
33. Ibid.
34. Ibid.
35. Xie Jinjin, "Career-Minded Chinese Send Russian to the Back of the Class," *The WorldPaper*, July 1996.
36. "Defining a Global Language," *The WorldPaper*, July 1996.
37. Ibid.
38. Ibid.
39. Diamond, op. cit. note 5.
40. Krauss, op. cit. note 2.

THE VITAL SIGNS SERIES

Some topics are included each year in Vital Signs; *others, particularly those in Part Two, are included only in certain years. The following is a list of the topics covered thus far in the series, with the year or years each appeared indicated in parentheses.*

Part One: KEY INDICATORS

FOOD TRENDS
Grain Production (1992–97)
Soybean Harvest (1992–97)
Meat Production (1992–97)
Fish Catch (1992–97)
Grain Stocks (1992–97)
Grain Used for Feed (1993, 1995–96)
Aquaculture (1994, 1996)

AGRICULTURAL RESOURCE TRENDS
Grain Area (1992–93, 1996–97)
Fertilizer Use (1992–97)
Irrigation (1992, 1994, 1996–97)
Grain Yield (1994–95)

ENERGY TRENDS
Oil Production (1992–96)
Wind Power (1992–97)
Nuclear Power (1992–97)

Solar Cell Production (1992–97)
Natural Gas (1992, 1994–96)
Energy Efficiency (1992)
Geothermal Power (1993, 1997)
Coal Use (1993–96)
Hydroelectric Power (1993)
Carbon Use (1993)
Compact Fluorescent Lamps (1993–96)
Fossil Fuel Use (1997)

ATMOSPHERIC TRENDS
CFC Production (1992–96)
Global Temperature (1992–97)
Carbon Emissions (1992, 1994–97)

ECONOMIC TRENDS
Global Economy (1992–97)
Third World Debt (1992, 1993, 1994, 1995)
International Trade (1993–96)
Steel Production (1993, 1996)

Part Two: SPECIAL FEATURES

Environmental Taxes (1996)
Private Finance in Third World (1996)
Storm Damages (1996)
Aid for Sustainable Development (1997)
Food Aid (1997)
R&D Expenditures (1997)
Urban Agriculture (1997)
Electric Cars (1997)
Arms Production (1997)

SOCIAL FEATURES
Income Distribution (1992, 1995, 1997)
Maternal Mortality (1992, 1997)
Access to Family Planning (1992)
Literacy (1993)
Fertility Rates (1993)
Traffic Accidents (1994)
Life Expectancy (1994)
Women in Politics (1995)
Computer Production and Use (1995)
Breast and Prostate Cancer (1995)
Homelessness (1995)
Hunger (1995)
Access to Safe Water (1995)

Infectious Diseases (1996)
Landmines (1996)
Violence Against Women (1996)
Voter Turnouts (1996)
Aging Populations (1997)
Noncommunicable Diseases (1997)
Extinction of Languages (1997)

MILITARY FEATURES
Nuclear Arsenal (1993)
U.N. Peacekeeping (1993)

Now you can import all the tables and graphs from *Vital Signs 1997* and all other Worldwatch publications into your spreadsheet program, presentation software, or word processor with the . . .

1997 WORLDWATCH DATABASE DISK

The Worldwatch Database Disk Subscription gives you current data from all Worldwatch publications, including the *State of the World* and *Vital Signs* annual book series, WORLD WATCH magazine, Worldwatch Papers, and Environmental Alert Series books.

Your subscription includes: a disk (IBM or Macintosh) with all current data and a FREE copy of *Vital Signs 1997*. In January 1998, you will receive a six-month update of the disk with a FREE copy of *State of the World 1998*. This disk will include updates of all long-term data series in *State of the World*, as well as new data from WORLD WATCH and all new Worldwatch Papers.

The disk covers trends from mid-century onward . . . much not readily available from other sources. All data are sourced, and are accurate, comprehensive, and up-to-date. Researchers, professors, reporters, and policy analysts use the disk to—

◆ *Design graphs to illustrate newspaper stories and policy reports*
◆ *Prepare overhead projections on trends for policy briefings, board meetings, and corporate presentations*
◆ *Create specific "what if?" scenarios for energy, population, or grain supply*
◆ *Overlay one trend onto another, to see how they relate*
◆ *Track long-term trends and discern new ones*

To order the 1997 Worldwatch Database Disk for just $89 plus $4 shipping and handling:
    Phone: (202) 452-1999 (credit cards accepted: Mastercard, Visa or American Express)
    Fax: (202) 296-7365; E-mail: wwpub@worldwatch.org; Website: http://www.worldwatch.org
    Or send your request to:

8813

1776 Massachusetts Ave., NW
Washington, DC 20036